D0934895

THE QUESTION OF ANIMAL AWARENESS

Evolutionary Continuity of Mental Experience

THE QUESTION OF ANIMAL AWARENESS

Evolutionary Continuity of Mental Experience

DONALD R. GRIFFIN

THE ROCKEFELLER UNIVERSITY PRESS

NEW YORK · 1976

COPYRIGHT © 1976 BY THE ROCKEFELLER UNIVERSITY PRESS

LIBRARY OF CONGRESS CATALOGUE CARD NO. 76-18492

ISBN 87470-020-5

PRINTED IN THE UNITED STATES OF AMERICA

SECOND PRINTING 1977

CONTENTS

Preface   vii

CHAPTER 1

Expanding Horizons in Ethology   3
> INTRODUCTION
> DEFINITIONS
> COMMUNICATION WITH ONE'S ENVIRONMENT

CHAPTER 2

The Versatility of Animal Communication   15
> SIGN LANGUAGE IN APES
> SYMBOLIC COMMUNICATION BETWEEN INSECTS
> SKEPTICISM ABOUT SYMBOLIC COMMUNICATION
>   BY HONEYBEES

CHAPTER 3

Is Man Language?   31

CHAPTER 4

Are Animals Aware of What They Are Doing?   39
> OPINIONS ABOUT AWARENESS IN ANIMALS
> EVIDENCE FROM NEUROPHYSIOLOGY
> DOES BEHAVIORAL COMPLEXITY IMPLY
>   CONSCIOUS AWARENESS?
> IS THERE EVOLUTIONARY CONTINUITY OF
>   MENTAL EXPERIENCES?
> THE NATURE AND NURTURE OF
>   MENTAL EXPERIENCES

CHAPTER 5

Objections and their Limitations   63

THE BEHAVIORISTIC OBJECTION
THE ANTHROPOMORPHIC OBJECTION
THE DANGERS OF RELAXING CRITICAL STANDARDS
THE "SO WHAT?" OBJECTION
AN OBSOLETE STRAIT JACKET

CHAPTER 6

The Adaptive Value of Conscious Awareness   81

CHAPTER 7

A Possible Window on the Minds of Animals   87

PARTICIPATORY INVESTIGATION OF
ANIMAL COMMUNICATION
TOWARD A COMPARATIVE LINGUISTICS

CHAPTER 8

Summary and Conclusions   103

Bibliography   107

Indexes   128

Preface

A FERMENT of constructive excitement is evident in ethology, the study of animal behavior with emphasis on evolutionary adaptations to the natural world. For example, social organization, individual recognition, altruistic behavior, endogenous activity rhythms or biological clocks, and complex systems of orientation and navigation have been identified in more and more species not previously suspected of having such complications in their ways of life. All scientific discoveries contain some element of novelty, but ethologists now feel confident in making statements that differ qualitatively from anything that was scientifically thinkable forty or fifty years ago. Since there is no reason to believe that this progress will suddenly come to a halt, it is worthwhile to outline some directions in which ethology may develop. In this attempt at speculative extrapolation, it is especially appropriate to pose some new questions and to reopen certain old ones from a fresh perspective. Most of these questions relate to the general issue of our evolutionary kinship to other species of animals, with special reference to the more complex cognitive functions that appear to regulate the behavior of animals and men.

Many thoughtful colleagues, individually and collectively, have provided essential stimulation without which this book would never have been written. Most important has been the stimulating environment of The Rockefeller University, which provided the opportunity for serious and consistent concentration on distant objectives. Thomas Nagel of Princeton University supplied an immediate spur while visiting our campus when he raised the question of whether animals have mental experiences. Peter Marler and Fernando Nottebohm, along with many other colleagues, offered invaluable and always constructive criticism. The "negative feedback" from my colleagues has been just as reinforcing as the positive encouragement. Rosanne Blair has patiently toiled over the preparation of innumerable draft manuscripts. Helene Jordan of The Rockefeller University Press has been a most perceptive editor. Finally, Jocelyn Crane supplied inestimable general encouragement, together with abundant factual knowledge from her wealth of experience with the real, natural world of animals and their ethology.

It is a pleasure to extend my grateful acknowledgments to the following publishers for permission to reproduce substantial quotations:

American Philosophical Association, *Thoughtless Brutes;* Braziller, *Signs, Language, and Behavior;* Clarendon Press, *Essays on Function and Evolution in Behaviour;* The Free Press, *Words and Things;* Harcourt Brace Jovanovich, *Language and Mind;* Harper & Row, *Cartesian Linguistics* and *Language: An Inquiry into its Meaning and Function;* Johns Hopkins Press, *Mind, an Essay on Human Feelings,* Vols. I and II; Macmillan, *Philosophical Investigations;* Oxford University Press, *Bowerbirds, Their Displays and Breeding Cycles;* Popular Library, *Supermoney;* Random House, *Psychological Explanation, An Introduction to the Philosophy of Psychology;* Teachers College Press, *Sciences, Psychology and Communication;* Wiley, *Learning Theory and Behavior* and *Learning Theory and Symbolic Processes;* Yale University Press, *The Philosophy of Symbolic Forms.*

I am also grateful to the following colleagues for permission to read and to quote from books or articles in preparation or in press: Colin G. Beer, Roger S. Fouts, James L. Gould, Peter Marler, Duane M. Rumbaugh, and W. John Smith.

THE QUESTION OF ANIMAL AWARENESS

Evolutionary Continuity of Mental Experience

CHAPTER

1

Expanding Horizons
in Ethology

INTRODUCTION

Ethologists and comparative psychologists have discovered
increasing complexities in animal behavior during the past few
decades. For example, food-finding, avoidance of predators, and
behavioral adaptations to environmental stresses by constructing
shelters, nests, and burrows all involve versatile tactics on the
animal's part. Social behavior, especially courtship and care of
developing young, call forth patterns of adaptive behavior
understandable only in terms of complex interactions of many
behavioral tendencies. Orientation, navigation, and communication behavior have provided especially striking examples of
previously unsuspected mechanisms at work. Although orientation and communication serve quite different functions, their
investigation has revealed similar surprises and extensions of
previous patterns of expectation. The flexibility and appropriateness of such behavior suggest not only that complex processes
occur within animal brains, but that these events may have much
in common with our own mental experiences. To the extent that
this line of thought proves to be valid, it will require modification
of currently accepted views of scientists concerning the relationship between animal and human behavior. Because of the

important implications of these developments in ethology, this book will examine both the pertinent evidence and its general significance in the hope of stimulating renewed interest in, and investigation of, the possibility that mental experiences occur in animals and have important effects on their behavior.

Before outlining what appear to be promising new approaches to these challenging questions, it will be necessary to discuss in some detail a number of theoretical considerations that have restricted scientific investigation of animal awareness. Because the available data are far from adequate, tentative speculations will often be necessary as first steps toward future investigations. This book will seek to raise or reopen significant questions, many of which have been neglected for some time, despite their basic importance. It will also try to relate such questions to previous discussions and to realistic possibilities for experimental investigations in the future. In discussions of these issues, we often find, in the absence of conclusive evidence, eloquent arguments and unqualified assertions. Because many of these can best be appreciated in the authors' own, carefully chosen words, a number of direct quotations are included, some in the text and others in annotations of certain bibliographical references. (The annotated references are marked in the text with °.) I believe that I have avoided the danger of distorting any writer's meaning by using quotations out of context, and any residual risk of such errors is more than offset by the great interest and significance of the questions at issue. In this way, readers can draw their own conclusions as directly as possible from the clearest statements of both the evidence and varying interpretations.

DEFINITIONS

Terms such as mental experiences, mind, awareness, intention, or consciousness are obviously difficult to define. But in the hope of clarifying the following discussion and minimizing misunderstandings, I will offer some rough-and-ready *un*sophisticated definitions which I believe will suffice at this very preliminary stage of raising questions and suggesting possible ways to explore a neglected area of ethology.

It is appropriate to begin with the most obvious fact about *mental experiences:* all of us have them. Every normal person thinks about objects and events that are remote in time and space from the immediate flux of sensations, and this is what I mean by mental experiences. A *mind* may be defined as something that has such experiences. *Awareness* is the whole set of interrelated mental images of the flow of events; they may be close at hand in time and space, like a toothache, or enormously remote, as in an astronomer's concept of stellar evolution. An *intention* involves mental images of future events in which the intender pictures himself as a participant and makes a choice as to which image he will try to bring to reality. Mental images obviously vary widely in the fidelity with which they represent the actual surrounding universe, but they exist in some form for any conscious organism. The presence of mental images, and their use by an animal to regulate its behavior, provide a pragmatic, working definition of *consciousness*.

It is important to emphasize at this point that the term consciousness is widely and strongly held by behavioral scientists to be useless for scientific analysis (Lashley, 1923; Boring, 1963; Hebb, 1974). One major reason for this rejection is the wide range of human mental experiences that intuitively seem to involve varying degrees and kinds of consciousness. (For a recent and stimulating discussion see "Smith", 1975.) Another reason is the unreliability of introspective verbal reports from human subjects concerning their mental experiences. I shall return to these thorny questions from several different viewpoints in Chapters 4, 5, and 7. For the time being, however, I propose to continue our inquiries on the basis of these working definitions, but with full recognition of the widespread reluctance of behavioral scientists to deal with the possibility that mental experiences occur in animals. Part of the difficulty may be the semantic uncertainty that results from "consciousness" being used to refer to so many different kinds of mental states. Any reader who finds it objectionably vague may "calibrate" it throughout this book by the working definitions presented above.

Another important point requiring careful attention is the obvious fact that our mental experiences include not only images and intentions, but also feelings, desires, hopes, fears, and a wide variety of sensations such as pain, hunger, rage, and affection. All such subjective entities have been largely rejected from traditional twentieth-century psychology and ethology for the same basic reason that they are "private data" directly observable only by the one who experiences them and describable to others only by introspective reports (Alston, 1972°). The qualities of sensations are especially troublesome and have almost wholly eluded objective definition. A classic example is that no one has been able to propose a satisfactory method to prove or disprove a statement such as "Your sensation of blue, stimulated by light of 450 nanometers, is exactly like what I report as red when my eyes are stimulated by 625 nanometer light. Each of us has learned to identify his private sensation according to the names given to light of these wavelengths or of objects that send such light to our eyes. Hence, agreement about nomenclature tells us nothing about the sensations actually experienced."

In attempting to come to grips with the question of possible mental experiences in animals, I shall concentrate on images, intentions, and awareness of objects and relationships in the outside world, rather than on feelings and purely subjective qualities. My reason for this choice is that I can see more realistic hopes of developing objective methods for gathering satisfactory data about the former than about what psychologists have called "raw feels" (Tolman, 1932). In Chapter 7, however, I shall return briefly to the question of subjective feelings in animals.

According to the working definitions presented above, it is not necessary to assume that consciousness or mental images are present only in living organisms; many have suggested computers as likely candidates (Scriven, 1963°; Apter, 1970°; Elithorn and

Jones, 1973; and Gregg, 1974). But the relationship between minds and computers is outside the scope of this book, which will confine itself to the possible existence of mental experiences and awareness in animals. It may be helpful, however, to draw an analogy between central nervous systems and the hardware of computers, while likening minds to the software, or programs. A difference of practical importance lies in the fact that each of us has his own mental experiences and thus, in a sense, sees his software from inside. The strict behaviorist insists on ignoring this source of information, on the ground that it cannot be observed directly by another person. But this may be as great a limitation as to observe a working computer and deny the existence of a program guiding its operation because the program cannot be seen on close inspection of the teletype, recording tape, or central processing unit.

Despite the self-consistent philosophical positions which deny that the nature and reality of mental experiences in other human beings can ever be demonstrated, I suggest that we accept the reality of our own mental experiences, even without rigorous proof. It certainly seems far more likely than not that mental experiences, whatever their actual nature may be, are closely linked to neurophysiological processes within our brains, even though we may not yet understand these processes at all well, and many important neural functions remain undiscovered. One possibility is that the relationship between mind and brain has important elements in common with the relationship between properties of whole animals and those of their constituent cells. The former, of course, depend utterly on the latter, but this does not mean that all important properties of the intact animal can be predicted from an understanding of its cellular physiology.

A helpful analogy is provided by considering a well-known and noncontroversial class of behavior—coordinated locomotion. No one supposes for a moment that an animal's walking, running, swimming, and flying require more than the activities of muscle cells, connective tissue, and neurons; no immaterial "essence of locomotion" is called for. But the patterns of structural and functional coordination by which thousands of cells produce bird flight, for example, are not easily derived from data on the endoplasmic reticulum or sliding filaments of actomyosin. If only for practical reasons, we are forced to deal with bird flight and similar examples of coordinated locomotion in terms appropriate for their level of organization; but this does not delude us into postulating vitalistic essences of flight independent of physics, chemistry, or cell biology. Although many details of locomotor physiology require further clarification, few, if any, scientists doubt that full explanations are possible in physiological terms. Nor need this confidence diminish our admiration for the success and beauty of the behavior under study. In some comparable fashion, I suspect, minds depend entirely on the functioning of central nervous systems (including neuroendocrine mechanisms), yet exhibit properties not easily predictable from even the most complete analysis of neurons and synapses. Similar concepts have been advanced by Lorenz (1963), Scriven (1963), K. Smith (1969°), Sperry (1969), Piaget (1971), Eccles (1973, 1974), Popper (1974°), and Hubbard (1975°), as well as by several contributors to a recent symposium edited by Ayala and Dobzhansky (1974).

COMMUNICATION WITH ONE'S ENVIRONMENT

After these introductory generalities, it is appropriate to come to grips with some specific cases. Orientation behavior provides a good starting point, because recent discoveries have shown that it is far more complex and versatile than scientists had anticipated. The orientation behavior of an animal can be viewed as a process of communication with its surroundings, in the sense that very weak signals from the environment trigger behavioral responses that release much greater physical energy and produce important biological results. Usually, but not always, pertinent information flows in only one direction—from the environment to the animal's sense organs. As in the case of true communication between individual animals, selective attention is paid to particularly useful physical signals, usually only one or a very few out of many available to the animal. In many cases, it has been difficult to identify the environmental signal that is actually selected, and a few specific examples illustrate how discoveries about orientation behavior have opened our eyes.

As a naive student in the late 1930s, I and others toyed with the notion that birds might orient themselves by the sun or stars during their homing or migratory flights. But my elders and betters were emphatic in discouraging what seemed to them— and, on more sober reflection, even to me—a rather silly and romantic line of speculation. "Why, the poor birds would need to carry around a whole set of tables, a sort of almanac, to correct for the motions of the sun and stars across the sky." No respectable biologist felt comfortable with anything more complex than orientation toward, away from, or perhaps at a fixed angle to a source of light. This viewpoint is well documented by Fraenkel and Gunn (1961). A sort of simplicity filter shielded us from worrying about possible complexities. In the 1950s, however, Matthews (reviewed in Matthews, 1968), Kramer (1959), and Sauer (1957) showed that birds are quite capable of making at least approximate corrections for the motion of the sun, and even of the stars, across the sky. Birds do indeed practice time-compensated sun- and star-orientation (Schmidt-Koenig, 1965; Emlen, 1967, 1975).

Similar simplicity filters held back my thinking for several years in another area where important progress was made at about the same time. Robert Galambos and I discovered echolocation in bats with the aid of then-unique electronic apparatus developed by the physicist G. W. Pierce (reviewed by Griffin, 1958). Pierce's apparatus was capable of detecting sounds above the frequency range of human hearing, and when I first brought bats to his laboratory it was obvious that ultrasonic sounds were emitted by these animals almost continuously. But when bats were allowed to fly near the apparatus, these sounds were only occasionally detectable. I did not realize at the time how directional Pierce's equipment was, that is, how greatly its sensitivity was reduced when bats were not close to the axis of the microphone. I therefore suspected that these newly discovered sounds might simply be call notes, not necessarily used for orientation. This conservative error was soon corrected after more detailed observations and experiments, but it is a significant example of the dangers inherent in limited imagination when one is dealing with new and unknown phenomena. Our experiments soon confirmed one of several speculative explanations, that of Hartridge (1920), for the ability of bats to fly without collisions through the complete darkness of caves by emitting sounds above the human frequency range and hearing echoes from obstacles.

Despite the excitement of solving this long-standing mystery, and despite the opportunity to measure the acoustical properties of the orientation sounds by which bats avoid obstacles (Griffin, 1946, 1950), it was several years before my thinking progressed to the point of seriously wondering whether bats might also use echolocation in hunting insects. Authoritative opinion warned that tiny flying insects would not return enough acoustical energy to yield audible echoes, and the whole idea simply seemed too farfetched for serious consideration. Yet here, again, the zoological reality turned out to exceed my initial speculations (Griffin, 1953, 1958). When I first took the trouble to observe wild bats hunting insects, it became clear that during insect pursuits they increase the repetition rate of their ultrasonic orientation sounds more sharply than had ever been observed in the laboratory. Such increases in repetition rate accompany the detection of small obstacles and also occur when a bat prepares to land. This was strong suggestive evidence that insects were detected by echolocation.

Conclusive experimental evidence was not obtainable for several years until we learned how to elicit insect-hunting behavior in relatively small rooms where controlled experiments were feasible (Griffin, Webster, and Michael, 1960). Little brown bats (*Myotis lucifugus*) learned to capture fruit flies at rates of several per minute. Weighing the bats before and after periods of intense feeding activity provided a direct measurement of hunting success. In a dark room filled with loud audio-frequency noise, which completely masked the faint flight sounds from the fruit flies, bats gained weight at essentially the same rate as they did in the quiet with lights on. In other experiments with relatively weak ultrasonic noise, the bats ceased all attempts to capture flying insects. More recent experiments have also demonstrated that bats have a highly refined capability for detecting faint echoes and also for discriminating among different classes of echoes according to their timing and frequency spectrum (reviewed by Simmons, Howell, and Suga, 1975). Echolocation by insectivorous bats is of sufficient importance that some groups of insects have even evolved auditory receptors sensitive enough to ultrasonic frequencies to warn them of approaching bats (Roeder and Treat, 1957; reviewed by Roeder, 1970).

Not only bats, but whales, porpoises, and dolphins use echolocation both for general orientation and to capture moving prey (reviewed by Kellogg, 1961; Norris, 1966; and Griffin, 1958, 1973). Two cave-dwelling birds find their way to nests that are sometimes deep inside caves where it is totally dark. Their echolocation, based on clicks that are clearly audible to human ears, suffices to detect obstacles as small as quarter-inch rods (Griffin and Suthers, 1970). A comparable surprise was the discovery that electric fishes orient themselves by sensing changes in the electric fields produced by their own electric organs (Lissmann, 1958; Lissmann and Machin, 1958). This initial discovery of a wholly unsuspected new sensory modality has been followed up by detailed investigations of the neurophysiology of electroreception, of the variety of signals used by different kinds of electric fish, and of the use of similar signals for social communication (reviewed by Bullock, 1973).

Still another unexpected discovery was the ability of honeybees not only to compensate for the motion of the sun through the sky, but also to orient themselves by the polarization patterns of the blue sky (Frisch, 1950). Sensitivity to polarized light has since been demonstrated in many other arthropods, and very recently in some individual homing pigeons by Kreithen and Keeton (1974a).

Important as all these discoveries were, they do not, of course, mean that all speculations about animal behavior will eventually prove to be correct. It may well be that, during the past generation, research on orientation behavior has disclosed a disproportionate share of such unexpected developments. Nevertheless, it is reasonable to inquire whether there is any end to this series of surprises. A contemporary case in point is the recent evidence that birds can sense and orient by the magnetic field of the earth. The positive evidence is limited so far to inconsistent and short-duration effects of magnets on initial headings of homing pigeons (Keeton, 1974; Walcott and Green, 1974) and slight shifts in the directional orientation of migrants in small cages (Wiltschko, 1974, 1975)—shifts so small that they are suggested only by statistical analysis of hundreds of responses. The situation is complicated by the fact that several attempts to demonstrate consistent and unequivocal responses of birds to weak magnetic fields have been totally unsuccessful (for example, Kreithen and Keeton, 1974b). On the other hand, stronger evidence for sensitivity to the earth's magnetic field has been reported in fish by Kalmijn (1971, 1974) and in honeybees by Lindauer and Martin (1972) and by Martin and Lindauer (1973).

Close study of previously unexplained capabilities for orientation has revealed the unsuspected sensory channels discussed above. In each case, a particularly appropriate physical signal from the environment is utilized selectively by a specialized sensory system. Echolocation and the electrical orientation of certain fishes are exceptional in being active processes during which the animal emits its own signals that yield useful information when conveyed back from the environment in the form of echoes or altered electric fields. An echolocating bat or porpoise literally questions its surroundings with orientation sounds that have been adapted for the function through a long evolutionary history. There are immediate and appropriate adjustments of both the motor mechanisms of sound emission and the neural networks of the auditory system to adapt the entire process of echolocation to such specific problems as detecting and avoiding obstacles, search for insect prey, and the pursuit and interception of flying insects. These reactions include increases in the repetition rate of orientation sounds when a difficult problem faces the bat; changes in the duration and pattern of frequency sweep used in each orientation sound; and changes in the sensitivity of peripheral portions of the auditory system within the bat's brain that occur a few thousandths of a second after each orientation sound is emitted. These neural adjustments are superbly adapted to enable bats to detect faint echoes returning from objects at short distances (reviewed by Simmons, Howell, and Suga, 1975).

The capabilities for perceptual organization which an animal requires for complex orientation behavior include the establishment of what has been called a cognitive map by Tolman (1948) or "an elementary map of the environment" by Thorpe (1974b). In discussing reductionism in biology, Thorpe points out that "problems of spatial position, of orientation and of direction finding . . . lead us naturally to the problem of conscious self-awareness." A suggestive example is provided by echolocating bats—not so much by their impressive successes as by situations in which their orientation fails. When flying through thoroughly familiar surroundings, many bats seem to rely heavily on spatial memory. Although orientation sounds continue to be emitted in an apparently normal manner, the bats collide with newly placed obstacles and turn back from the former location of objects that have suddenly been removed (Moehres and von Oettingen-Spielberg, 1949; Griffin, 1958). Bats behaving in this way remind me of the collision of the *Andrea Doria* and the *Stockholm* in thick fog when, according to contemporary newspaper accounts, both ships were equipped with properly functioning radar sets. Thus I like to call bats which are relying so heavily on spatial memory or internal maps "Andrea Doria bats." Under these conditions, apparently they pay attention only to their internal images of spatial relationships, even though echoes from the newly placed obstacle reach their ears at far higher intensity levels than those of small objects or insect prey which they detect readily under other circumstances. In any event, this kind of behavior clearly demonstrates that some sort of internal map or stored pattern representing the familiar environment must exist in the bat's brain.

While it is quite possible to analyze patterns of animal behavior which suggest the presence of a detailed internal map without inferring that the animal is consciously aware of any aspects of the geometry of its surroundings, it may be more conservative to recognize that, in fact, we simply do not know. The evidence that animals employ some sort of internal imagery of their surroundings suggests a need to reconsider the general question of subjective mental experiences in animals. Renewed attention to these long-neglected but highly significant questions is now especially appropriate and timely, because recent advances in ethology have opened up new lines of experimentation that offer realistic hopes of answering them in due course. As suggested in rather general terms in Chapter 7, it even seems likely that we can anticipate the eventual emergence of a truly experimental science dealing with the mental experiences of other species.

While the orientation and navigation of certain animals provide many examples consistent with the assumption that the animal has an internal image of its surroundings, even more pertinent evidence bearing on these general questions is provided by recent discoveries about communication behavior as it is practiced by several species.

CHAPTER

2

The Versatility of
Animal Communication

THE ANALYSIS OF COMMUNICATION between individual animals has led to several discoveries of the highest significance. From investigations of a wide variety of species belonging to several phyla, from fiddler crabs (Crane, 1975) to chimpanzees (Gardner and Gardner, 1971), a common thread of versatile diversity can be discerned. Although something simpler was initially expected, communication signals have turned out, at the very least, to include an announcement that the sender is of a given species, sex, and appropriate age, and is in one of a relatively few basic behavioral states, such as readiness for fighting, fleeing, or mating (Sebeok and Ramsay, 1969; Hinde, 1972; Smith, in press). These messages also have an intensity scale from weak to strong. Conspecific partners respond to varying degrees and in different ways, but often appropriately according to their own age or reproductive condition. Individual recognition of conspecific companions is common at least in birds and mammals (Falls, 1969; Beer, 1973a, 1973b, 1975, 1976). A frequent element is the flexibility and interrelatedness of the signaling behavior; fairly complex sequences are performed, with each step depending on an appropriate signal or response from the partner.

Almost every sensory system is employed by some species of animals for communication with conspecifics. Chemical signals, including pheromones, are ordinarily detected by the olfactory system and are especially important in insects, flying phalangers, rodents, cats, and monkeys (Wilson, 1975). Sounds are extensively used by many groups of invertebrates, as well as by all classes of vertebrate animals (Sebeok, 1968, 1972). Surface waves are used by aquatic insects (Wilcox, 1972). Tactile communication includes not only direct contact between animals, but communication via vibrations of the ground or vegetation. Leaf-cutter ants stridulate when accidentally buried, and other members of the colony locate them by vibrations transmitted through the soil (Wilson, 1971). In certain spiders, the male begins his courtship by setting the female's web into a particular pattern of vibrations. Many groups of fishes that use electrical orientation (Bullock, 1973) also communicate by electrical signaling (Hopkins, 1974; Westby, 1974). Communication by visual signals is widespread (Marler, 1968). An especially striking example is the courtship of certain fireflies, which exchange light flashes signaling sexual readiness (Lloyd, 1966, 1975). But visual signaling has not been studied as extensively as has acoustical communication, primarily because it is technically more difficult to record and play back visual signals.

SIGN LANGUAGE IN APES

The recent studies of gestural communication between chimpanzees and human experimenters are widely recognized as a major breakthrough in the behavioral sciences (Gardner and Gardner, 1969, 1971, 1975°). Several earlier attempts to teach chimpanzees to make vocal sounds were significant in their almost total failure. Even after years of effort, home-reared chimpanzees learned to produce only a very few recognizable monosyllabic words, although they recognized many words of human speech. The Gardners, stimulated in part by Goodall's (1968) observations of wild chimpanzees, decided that gestures were a more promising method for communication. They trained a wild-born young female chimpanzee, Washoe, to use several dozen "words" from the American Sign Language for the deaf. An important part of their procedure was the total immersion of Washoe in a social environment consisting of human companions who communicated only in this sign language while in her presence. In four years, Washoe acquired approximately 130 signs, invented a few of her own, and used them all in conversational exchanges with her human companions. In carefully controlled blind experiments, she was able to name pictures presented by an experimenter who could not see them himself. Washoe spontaneously used signs and combined small groups of signs in meaningful ways, transferring them appropriately to new situations. For example, the sign for "open," which she originally learned for doors, she later used to request the opening of boxes, drawers, briefcases, and picture books.

Washoe learned to use gestural signals much as words are used by young children, but of course many differences remain between her signing and early human speech. For example, word order seems to play a smaller role in Washoe's signing than it does with children who have vocabularies of comparable size. Investigations of gestural communication by chimpanzees have been continued both by the Gardners and by Fouts, Lemmon, and their colleagues at the University of Oklahoma (reviewed by Fouts, 1973, 1975) and Fouts and Rigby (in press). Among many significant findings, these studies have demonstrated that chimpanzees can communicate with each other by means of a sign language they have been taught by human experimenters. They can also learn to identify objects and pictures on hearing the names in spoken English. This ability allowed Fouts, Chown, and Goodwin (in press) to train a chimpanzee to utilize both spoken English and sign language. A three-year-old male chimpanzee, Ally, acquired a vocabulary of more than 70 reliable signs and also learned to understand several spoken phrases and words. He was then taught new signs corresponding to 10 spoken words to which he was already responding correctly. These were names of familiar objects, but the objects were not present during this phase of the training. After training was completed, Ally showed himself completely capable of using these gestural signs correctly to identify the objects for which they stood.

Premack (1971) and others (Rumbaugh et al., 1974; and Gill and Rumbaugh, 1974) have studied the languagelike behavior of chimpanzees by different types of experiments. These utilize a relatively small number of symbolic objects or mechanical devices which the chimpanzees learned to use appropriately. In Premack's experiments with Sarah, colored plastic tokens were used as names for familiar objects, and Sarah learned to use these tokens to request specific items of food. In the more recent experiments by Rumbaugh and his colleagues, a computer keyboard is used correctly by the chimpanzee to request desired objects or simple actions. In these experiments, the vocabulary is limited to symbols or keys provided by the experimenter, and although relatively large repertoires have been built up and used appropriately, the experimental situation precludes or greatly limits the possibilities for the animal to acquire a large vocabulary and to generate new "words" spontaneously. These experiments can be more rigorously controlled, but they offer the chimpanzee less scope for originality than do the methods used by the Gardners and by Fouts. These differences in experimental approach are less important than the fact that both approaches have yielded similar results: chimpanzees have learned to use surprisingly large vocabularies of gestures or manually manipulated symbols to communicate far more complex messages than scientists had previously believed were possible in any nonhuman animal.

The details of languagelike communication learned by chimpanzees have been discussed and reviewed extensively, for example by Klima and Bellugi (1973), Linden (1974), Thorpe (1974a, 1974b), Bronowski and Bellugi (1970°), S. J. Gould (1975), Griffin (in press), and Fouts and Rigby (in press). One interpretation is that the basic ability to communicate is severely limited by the anatomy and physiology of the chimpanzee larynx. This seems less likely than the alternate hypotheses that the chimpanzee brain is capable of relatively complex communication but that this capability can be expressed far more readily through manual gestures than by vocalization. The extensive observations by Goodall (1968, 1971, 1975) have clearly demonstrated that wild chimpanzees use gestures and facial expressions that are effective, but are difficult for human observers to analyze in detail. Furthermore, Menzel (1974) and Menzel and Halperin (1975) have shown that captive chimpanzees can communicate fairly complex information by some combination of gestures or expressive movements that human investigators have not yet deciphered. [figured out] In these experiments, several chimpanzees who were familiar with each other and with a large outdoor enclosure were confined temporarily in small cages at one edge of this enclosure. Then one animal was led to something such as food, not visible from any of the isolation cages, shown the object, and returned to the isolation cage. Next, the entire group was released, and the "leader" was able to convey the location of the hidden object rapidly and efficiently. Sometimes, when the leader seemed not to wish his companions to discover the object, he appeared to attempt to keep them from locating it. These apes seem capable of intentionally conveying or withholding information from their companions. If these experiments had been conducted with silent human beings, an observer would have had no doubt that the leader knew where the object was located and either did or did not wish his companions to find it.

SYMBOLIC COMMUNICATION BETWEEN INSECTS

In many ways, an even more surprising discovery about animal communication is the *Tanzsprache* (literally "dance speech") of honeybees. Our understanding of this flexible communication system in a highly social insect is based on the brilliantly pioneering experiments and insights of Karl von Frisch (1923,

1946, 1967, 1972, 1974). The communicative dances of honeybees take several forms, but the most significant is the *Schwanzeltanz* (usually translated "waggle dance"), which is a figure-eight-shaped pattern ordinarily carried out inside a hive in total darkness by bees crawling rapidly about over the vertical surface of the honeycomb. The most common situation in which bees execute these waggle dances is when a forager has returned from a rich source of food and carries either nectar from flowers in her stomach or pollen grains packed into basketlike spaces formed between specialized hairs on her legs. One cycle of the waggle dance consists of a circle with a diameter about three times the length of a bee, followed by a straight portion and then another circle turning in the opposite direction from the first, after which the straight segment is repeated. The circling thus alternates clockwise and counterclockwise. The straight portion is the important component for transferring information, and it is during this part of the figure-eight pattern that the abdomen is moved vigorously from side to side at 13 to 15 times per second.

Although these waggle dances had been observed for hundreds of years, it was von Frisch who first noticed that they varied in a meaningful fashion. He was studying bees in an observation hive—a small, thin structure with space for only a single layer of honeycomb, but provided with a glass window on one or both sides so that the bees can be seen while carrying out all the many activities involved in maintenance of the hive and care of the queen, eggs, and developing larvae. Von Frisch was feeding his bees concentrated sugar solutions in small dishes, and for convenience these dishes were placed close to the hive. The bees were marked individually with small daubs of paint while they filled their stomachs at the feeding stations. These bees danced on returning to the observation hive by moving only in circles— called round dances—alternately clockwise and counterclockwise, but other bees from the same hive were bringing pollen back from distant flowers. Von Frisch noticed that the pollen-carriers were executing waggle dances, and concluded that the two patterns were somehow correlated with the type of food material. Not until more than 20 years later did he happen to study the dances performed by bees that had brought back the same type of food from different distances. The waggle dances now occurred when the food was more than approximately 100 meters from the hive, showing that the primary factor was distance, rather than type of food.

Von Frisch followed up this observation with an extensive series of experiments which demonstrated that, whenever bees perform waggle dances after gathering food at distances beyond 100 meters, the straight waggling portion of the dance varies in length and direction, and that these variations are closely correlated with the location of the food source. Direction is indicated relative to the vertical, inside the hive, and to the location of the food relative to the position of the sun in the sky. Thus, if food is directly toward the sun, the straight waggling run is oriented straight up; if directly away from the sun, straight down; and correspondingly at intermediate directions. The length and duration of the waggling run correlates well with the distance from hive to food. Hence, a human observer who knows the "code" discovered by von Frisch can observe the waggle dances and "read" them well enough to determine the location of the food with an accuracy of approximately ±5 or 10 degrees and ±10 per cent in distance. A series of these dances specifies quantitatively the distance and direction, and qualitatively the desirability, of what a scout bee has located (the sugar concentration of nectar, for instance).

Although directions are ordinarily expressed relative to the vertical, under special conditions bees carry out the same type of dance on a horizontal surface in view of the sky, and the waggle run then points directly toward the food. The dances are most frequently used to signal the location of a food source, but under special conditions they are also applied to other requirements of the mutually interdependent colony of bees. They are not rigidly used after all foraging flights. When food is plentiful, returning foragers often do not dance at all. The odors conveyed from one bee to another always help to direct recruits to new food sources, and often they alone are sufficient. Independent searching by individual foragers seems to be adequate under many conditions. Thus, the dance communication system is called into play primarily when the colony of bees is in great need of food. But it is not tightly linked to any one requirement. Lindauer (1955) showed that the same dances are used for such different things as food, water, and resinous materials from plants (propolis). Moreover, when a colony of bees is engaged in swarming, scouts search for cavities suitable to serve as a future home for the entire colony and report their location by the same dances, which are now performed by crawling over the mass of bees that makes up the swarm cluster (reviewed by von Frisch, 1967, and Lindauer, 1971a).

Odors from flowers or scented secretions by the bees themselves add specific information to the dance pattern, and such odors enable the bees to locate the exact position of the food or other desiderata. The dancer conveys information as to what she is dancing about by transferring odors to other bees which cluster around her. Sometimes this transfer is the result of direct contact between odorous particles adhering to her external surface and the antennae of other bees. Odors are also transferred along with the stomach contents that are regurgitated by the returning forager and eagerly sucked up by the other bees, which pay close attention to the dancer. These odors from flowers or from scented secretion are apparently used to locate the source once the bee has reached its vicinity. Sounds or vibrations are also involved, at least in conveying the arousal level of the dancing bee (Esch, 1961; Wenner, 1962).

Whereas human observers can see the patterns of honeybee dances through the window of an observation hive, the bees do most of their dancing in darkness, because the entrance to an ordinary hive is very small and virtually no light can penetrate into the labyrinthine recesses between multiple layers of honeycomb. Tactile sense organs (mainly those of the antennae and in the joints of the exoskeleton) must be the primary channels by which the dancer transmits information through her waggle runs to follower bees, who can somehow relate these complex jostlings to the force of gravity. Later, the follower must also be able to orient her flight at the same angle to the sun as the dancer did to gravity during the waggle dance.

When Lindauer observed scouts from a swarm of bees that had moved only a short distance away from the original colony, he found that the same marked bee would sometimes change her dance pattern from that indicating the location of a moderately suitable cavity to one signaling a better potential site for a new hive. This occurred after the dancer had received information from another bee, and had flown out to inspect the superior cavity. Thus, the same worker bee can be both a transmitter and receiver of information within a short period of time and, despite her motivation to dance about one location, she can also be influenced by the similar, but more intense, communication of another dancer. Despite the profound implications of these investigations, the practical difficulties of controlled experiments with swarming bees have so far prevented others from extending them. But there is no escape from the conclusion that, in the special situation when swarming bees are in serious need of a new location in which the colony can continue its existence, the bees exchange information about the location and suitability of potential hive location. Individual worker bees are swayed by this information to the extent that, after inspection of alternate locations, they change their preference and dance for the superior place rather than for the one they first discovered. Only after many hours of such exchanges of information, involving dozens of bees, and only when the dances of virtually all the scouts indicate the same hive site, does the swarm as a whole fly off to it (reviewed by Lindauer, 1971a). This consensus results from communicative interactions between individual bees which alternately "speak" and "listen." But this impressive analogy to human linguistic exchanges is not even mentioned by most behavioral scientists, for instance Brown (1975), who devotes a whole chapter in his excellent textbook to the dances of bees.

Honeybees carry out several other types of dancelike motions, some of which appear, from very limited evidence, to have a communicative function (Frisch, 1967, pp. 278–284). One of these is the *Schwirrlauf*, or buzzing run, performed on the surface of a swarm after the dancing scouts have reached a nearly complete consensus about a particular location. It seems to convey a sort of imperative "Let's go!" and is followed by a mass flight by the entire swarm. Details of the waggle dances may also convey information other than the geometrical patterns related to direction and distance. More information may be conveyed by these less-conspicuous motions, and hence it would be premature to conclude that we understand completely the communication behavior of honeybees.

Having received the facts of the situation, it is appropriate to turn briefly to some questions of semantics. Many will object to calling bee dances symbolic, although they meet one of the meanings of "symbol" in the Random House *Dictionary of the English Language:* "Something used for or regarded as representing something else," provided that we may consider gestures and motions to be *things* in the sense of this definition. Charles Morris (1946°) proposed a special meaning for the term symbol, one that is applicable to bee dances. He first defined a sign as "something that directs behavior with respect to something that is not at the moment a stimulus," and proposed that a symbol be defined as "a sign produced by its interpreter which acts as a substitute for some other sign with which it is synonymous; all signs not symbols are signals."

In the case of bees ready to seek food, the waggle dance substitutes for direct leading of recruits to the food or pointing directly to it. Direct pointing does occasionally occur when honeybees (*Apis mellifera*) dance on horizontal surfaces, and this sort of dance is the only type employed by the dwarf honeybee, *Apis florea* (Lindauer, 1971a). Such directly pointing dances might be considered signs in Morris's terms, whereas dances transposed to gravity as a directional reference would qualify as symbols. Alternately, even the dances on horizontal surfaces fulfill Morris's definition of a sign about another sign, specifically about the process of actually leading the way to the food. Thus they, too, might qualify as symbols. On this interpretation, the transposed dances carried out on vertical surfaces in the dark hive would qualify doubly as symbols. Certain ants (including *Cardiocondyla venustula* and members of the genus *Camponotus*) do lead members of their colony to food by a process called tandem running, in which one ant grasps another and drags or leads it toward a newly located source of food (Wilson, 1971).

Regardless of the semantic issues discussed above, it is clear that the versatility of honeybee dances raises basic questions for which we had been poorly prepared by the behavioristic tradition in psychology or the comparable reductionism in biology. Such complex signaling would have been surprising enough in mammals. Suppose, for example, that a comparable form of symbolic gesturing, pointing the way to any of several sources of a variety of desiderata had been discovered in some previously unstudied monkey. What an impact such a finding would have had on theories about the origins of human language! To find symbolic communication in an insect was truly revolutionary. If an insect brain weighing only a few milligrams can manage flexible two-way communication, the possibility clearly arises that languagelike behavior may occur in other animals, as well. In other words, the occurrence of symbolic communication in two groups as distantly related as Hymenoptera and Primates (whose evolutionary lines of descent diverged at least 500 million years ago) suggests that such behavior is not the exclusive prerogative of any one species.

Another example may be provided by the report of Chauvin-Muckensturm (1974) that great spotted woodpeckers (*Dendrocopos major*) can learn a simple telegraphic drumming code by which they request whichever of five types of food they wish to obtain from the experimenter at a particular time. When a woodpecker had learned to use this drumming code to communicate with the experimenter, other persons with whom the bird was not familiar were able to communicate with approximately equal effectiveness. These controls do not entirely eliminate the possibility of a "Clever Hans error," named for a horse, Clever Hans, which apparently had learned to carry out complex arithmetical communications and transmit the results by tapping with one foot. Although many scientists were convinced of the genuineness of this accomplishment, more careful study showed that the horse was actually watching the person who presented the problem by writing numbers on a blackboard, and who had to count in order to determine whether the horse was giving the correct answer (Pfungst, 1911). In the course of such counting, small movements were performed unconsciously, and it was these which Clever Hans had learned to notice. He had also learned to stop tapping when the experimenter stopped nodding or otherwise indicating his own process of counting. Because new human observers also gestured in minor ways during their counting, without realizing that they were doing so, Clever Hans was able to perform his apparent feats of arithmetic even for strangers. Despite these limitations, the experiments of Chauvin-Muckensturm are, at the very least, suggestive and deserve to be extended and elaborated.

Skepticism about Symbolic Communication by Honeybees

The discovery that insects employ a communication system that edges so close to human speech in its symbolism and flexibility has far-reaching implications which may well have played a part in the skepticism expressed by Adrian Wenner (1971, 1974), Wells (1973°), Wells and Wenner (1973), and doubtless felt by many others, such as Hinde (1970), Langer (1972°), Tavolga (1974), or Glucksberg and Danks (1975°). Wenner and his colleagues have seriously questioned whether the evidence presented by von Frisch and Lindauer really does suffice to demonstrate communication of information about distance and direction. They contend that site-specific odors can account for the results of von Frisch's experiments, and that bees do not convey to one another information about distance and direction. They concede that the dance patterns are closely correlated with the distance and direction of a food source from which the dancer has just returned, but interpret this as a sort of accidental epiphenomenon. As pointed out by Gould (1976), correlation between a behavior pattern and some other process does not prove that the behavior serves for communication. But this line of skeptical thought leaves one in the awkward position of having no explanation to offer for the remarkably specific correspondence of dance pattern with the appropriate distance and direction.

Wenner underemphasizes the extent to which von Frisch, many years earlier, had described extensive experiments which showed that odors are of great importance in recruiting bees to new food sources. He sets up a sort of straw man by implying that von Frisch claimed that bees *always* dance or that they locate food *only* by information conveyed by the dances. Because bees often locate food by odor, Wells and Wenner deny that the dances convey information about distance and direction under any circumstances. Yet their unwarranted skepticism has had the constructive effect of stimulating several new and improved experiments (Esch and Bastian, 1970; Gould, Henerey, and MacLeod, 1970; and Lindauer, 1971b). The crux of the issue clearly lies in the behavior of bees stimulated by the dances. Wenner and his associates did point out weaknesses in the experiments by which von Frisch had tested the degree to which recruits actually fly to the location indicated by the dances. Many of the results of these experiments could be explained as searching for odors conveyed during the dance.

J. L. Gould (1974, 1975a, 1975b, 1976) has confirmed as conclusively as seems at all feasible that information about direction and distance is indeed conveyed by the dances. He devised a procedure that allows him to alter the direction of the dance so that it describes a different location from that of the actual food source. To do this, Gould took advantage of two details of honeybee behavior that had previously been discovered by others. The first is that if a bright, concentrated light is provided near an observation hive so situated that the bees do not have a direct view of the sky, the bees may interpret this light as though it were the sun and orient the dances relative to it, rather than to gravity. Thus, if the food were located 90° to the right of the sun, the dances would ordinarily be pointed 90° to the right of straight up. If the observation hive is inside a small building without windows, and if a bright light is placed to the left of the hive, the bees often orient their dances 90° to the right of this light or approximately upward. Ordinarily, such an artificial light does not alter the efficiency of the dance communication, because both dancers and potential recruits are affected in the same way. The dances are therefore interpreted correctly, even though the reference point is shifted from gravity to the artificial light.

The second detail of bee behavior utilized by Gould is the use of the ocelli—small eyes near the top of the head between the large compound eyes—to monitor the general level of illumination. If the ocelli are covered with opaque paint, the bees behave as though the light level were much lower than it actually is, even though their compound eyes are intact. In the special situation where a bright light inside a shed containing an observation hive causes a reorientation of the dances, covering the ocelli with black paint can prevent this reorientation to the artificial light.

In Gould's experiments, the ocelli of foragers were painted with black paint at the feeding station, and an appropriate artificial light was provided near the observation hive where they danced. By carefully adjusting the position and brightness of the light, he was able to avoid reorienting the dances, but shifted the reference point for untreated recruits. The net effect was that the dance communication system could be distorted experimentally so that the dancers pointed toward a location different from the actual food source from which they had just returned. The great majority of recruits flew to the one of several test feeders at the distance and direction indicated by the experimentally altered dance, and not to the place from which the dancer had actually returned. One should not underestimate the ingenuity with which sufficiently determined skeptics can find some tortuous loophole providing for an indirect effect of odors, but the weight of evidence is now overwhelmingly in favor of von Frisch's original interpretation. For independent reviews of these questions, see Wilson (1971) and Michener (1974).

A second sort of skepticism about the "dance language" of honeybees takes a quite different form from Wenner's. For example, Smith (1968 and in press) and Langer (1972) accept the conclusion that bees obtain information about distance and direction from the waggle dance, but they interpret the dance not as actual communication about external objects, but rather as a function of the inner state of the dancer. On this basis, Langer rejects any conclusion that information transfer by dancing bees is at all comparable to human communication. This and similar opinions will be further analyzed in Chapters 4 and 5. But this analysis will be facilitated by first considering in the next chapter certain basic ideas about the nature of human language and its relation to human thinking.

3

Is Man Language?

HUMAN LANGUAGE is often contrasted with all kinds of communication behavior in other species, and the distinction is held to be *the* primary difference in kind that distinguishes human beings from animals (Bloomfield, 1933°; Monod, 1975°). Many philosophers and linguists have also argued that human language is closely linked with thinking, if not identical and inseparable (Cassirer, 1953°; Fodor, Bever, and Garrett, 1974; Hattiangadi, 1973; Healy, 1971; Lenneberg, 1971; Pyles, 1971; Thass-Thienemann, 1968; and Weis, 1975). Langer (1942, 1962°, 1967, 1972°) expounds this view with special eloquence and vigor.

Black (1968) assures us that "It would be astounding to discover insects or fish, birds or monkeys, able to *talk to one another* . . . [because] . . . Man is the only animal that can talk . . . that can use *symbols* . . . the only animal that can truly *understand* and *misunderstand*. On this essential skill depends everything that we call civilization. Without it, imagination, thought—even self-knowledge—are impossible." The neurologist Critchley (1960) was so impressed by human speech that he wondered: "Can it be, therefore, that a veritable Rubicon does exist between animals and man after all? . . . Can it be that Darwin was in error when he regarded the differences between man and animals as differences merely in degree?" Goldstein (1957) asserted in the same vein that "Language is an expression of man's very nature and his basic capacity. . . . Animals cannot have language because they lack this capacity. If they had it, they would . . . no longer be animals. They would be human beings." To Anshen (1957), "Man *is* language."

Noam Chomsky is an influential contemporary philosopher who, in his penetrating discussions of the nature of language (Chomsky, 1966, 1972), subscribes to the tradition begun by Descartes. Both Descartes and Chomsky discern a creative property of language and frequently assert, or at least imply, that animals are merely machines, whereas language is the essence of humanity. In a blend of translation and reiteration, Chomsky (1966) summarizes the Cartesian view that no men are "so depraved and stupid, without even excepting idiots, that they cannot arrange different words together, forming of them a statement by which they make known their thoughts; while, on the other hand, there is no other animal, however perfect and fortunately circumstanced it may be, which can do the same . . . man has a species-specific capacity, a unique type of intellectual organization which cannot be attributed to peripheral organs or related to general intelligence and which manifests itself in what we may refer to as the 'creative aspect' of ordinary language use—its property of being both unbounded in scope and stimulus-free. . . . Human reason, in fact, is a universal instrument which can serve for all contingencies, whereas the organs of an animal or machine have need of some special adaptation for any particular action . . . no brute [is] so perfect that it has made use of a sign to inform other animals of something which had no relation to their passions . . . for the word is the sole sign and the only certain mark of the presence of thought hidden and wrapped up in the body; now all men . . . make use of signs, whereas the brutes never do anything of the kind; which may be taken for the true distinction between man and brute."

Chomsky continues, "The unboundedness of human speech, as an expression of limitless thought, is an entirely different matter [from animal communication], because of the freedom from stimulus control and the appropriateness to new situations. Modern studies of animal communication so far offer no counterevidence to the Cartesian assumption that human language is based on an entirely different principle. Each known animal communication system either consists of a fixed number of signals, each associated with a specific range of eliciting conditions or internal states, or a fixed number of 'linguistic dimensions', each associated with a non-linguistic dimension." The evidence reviewed in Chapter 2 calls into serious question these sweeping, negative generalizations of Descartes and Chomsky.

Price (1938) argued that if animals use symbols, we must assume they have minds. Bee dances are certainly symbolic, but Chomsky (1972) argues that one cannot trace similarities and evolutionary continuities between animal and human communication. "When we ask what human language is, we find no striking similarity to animal communication systems ... human language, it appears, is based on entirely different principles. This, I think, is an important point, often overlooked by those who approach human language as a natural, biological phenomenon; in particular, it seems pointless, for these reasons, to speculate about the evolution of human language from simpler systems. ... As far as we know, possession of human language is associated with a specific type of mental organization, not simply a higher degree of intelligence. There seems to be no substance to the view that human language is simply a more complex instance of something to be found elsewhere in the animal world."

Cultural transmission of human language has often been cited as one criterion establishing it as unique to our species. For example, Pollio (1974°) states three criteria necessary to qualify an event as a symbol: it must be representative of some other event, "freely created," and transmitted by culture. The dances of honeybees are recognized as being representative, but are held to be too rigid and unvarying to satisfy the second criterion, and to be genetically programmed rather than culturally transmitted.

Yet, under the same conditions, the waggle dances do vary some-what, and their performance does not necessarily occur whenever a bee returns to the hive from a source of food (Frisch, 1967). There is no direct evidence for cultural transmission of bee dances, but the development of individual worker bees is so heavily dependent upon care provided by other members of the colony that cultural influences are difficult to exclude. The specific dance pattern used for a given food source is, of course, learned by the individual worker during her foraging flights. Thus, the genetic instructions involve the ability to learn a coded pattern of communication behavior. Therefore, arguments similar to those of Chomsky concerning the probable genetic basis for human linguistic ability could be made concerning honeybees. Lorenz (1969°) has reviewed the considerable evidence that cultural transmission is important in the social behavior and communication of birds. Sarles (1975) has reviewed the difficulty of basing a rigid human–animal dichotomy on the criterion of language.

Of course, no one in his senses can overlook the enormous difference in complexity, subtlety, and versatility that separates human language from any known, or even speculatively postulated, communication between members of other species. But most scholars and scientists concerned with the question have not been content with quantitative distinctions, differences in degree rather than differences in kind. Thus Hockett (1958), Hockett and Altmann (1968), and Thorpe (1972, 1974a, 1974b) have tried to formulate objective criteria by which human language can be qualitatively distinguished from animal communication. Although the task seems to become increasingly difficult as more and more is learned about communication in other species, it is important to review the 16 design features tabulated by Thorpe (1974a). All of these features are certainly present in human language, and the question that arises is the degree to which any of them, or any combination, provide an objective basis for concluding that there is a fundamental difference in kind between human language and all communication systems used by other animals.

Some of the 16 design features are of insufficient basic importance to warrant setting them up as criteria for a fundamental difference in kind. Others are, in fact, present in many animal-communication systems. These include *reliance on the vocal-auditory channel; broadcast transmission and directional reception; rapid fading; interchangeability* (animals can act both as transmitters and receivers); *specialization* (energy in the signal small compared to the effects triggered by it); and *complete feedback* (transmitting animal able to perceive all relevant properties of his signal). Another set of design features seems, at first thought, to be distinctively human, but similarities are certainly present in many animals. These include *semanticity*, defined as use of signals to correlate and organize the activities of a community on the basis of associations between the signals and properties of the surrounding world. Many animal communication signals certainly satisfy this criterion in a general way. For example, territorial songs of birds correlate in an important fashion with the properties of the environment as far as conspecifics are concerned. *Arbitrariness* is another criterion that falls into this category. Bee dances are often considered not arbitrary because the dance pattern is a sort of iconic replica of the route to be flown, but there are so many other aspects of the dances, such as their vigor and the role of sounds in conveying something akin to the urgency of the message, that it becomes a matter of semantics whether to designate these features as arbitrary.

Eight other design features are more difficult to find outside of human language. *Discreteness* is an important property of human linguistic communication, in that small elements, such as words or syllables, do not functionally grade into one another. But the definition of discreteness depends heavily upon the size of element considered. For example, a single cycle of the honeybee dance or even a single cycle of abdomen waggling could well be considered a discrete unit. The latter, in particular, is combined in various ways with other elements, such as sound pulses. Two other features—*tradition*, the meaning of signals transmitted by teaching and learning, and *learnability*, users of the communication system learning about it from one another—are closely related and perhaps should be considered together. It is clear that learning and social tradition play a large role in the details of bird song. Bee dances are generally considered to be genetically programed, but here, too, the details are certainly learned, as when bees visit and dance about a location conveyed to them by other dancers.

Another design feature frequently stated to be lacking in animal communication systems is *duality*. A system is said to have duality if signal elements are meaningless in themselves but become meaningful when formed into appropriate combinations. Here, again, the applicability of the criterion depends upon the size of unit considered. Bee dances or other forms of communication behavior can easily be subdivided into individual elements, such as single muscle contractions, which by themselves would have no communicative significance. Human language obviously achieves a great deal of its enormous importance by use of compound and complex combinations of small elements, but we do not know enough about animal communication to judge the degree to which combinations, as opposed to individual signals, may be important.

One design feature often claimed to be unique to human language is *displacement*. Displacement means that the communication process can refer to things remote in time or space. Clearly, bee dances satisfy this criterion, and so do some communication signals used by other animals, such as alarm calls announcing that a flying predator is overhead. Another similar criterion is *openness*, meaning the ease and frequency with which new messages are coined by using previously unused combinations of elements of the communication system. This is sometimes also called *productivity*. Yet it has been obvious ever since von Frisch's first reports about bee dances that these insects frequently dance concerning locations and kinds of food about which they never danced before. A final criterion of *reflectiveness*, or the ability to communciate about the communication system itself, is a relatively recent addition to the list. Thorpe feels this property "is undoubtedly peculiar to human speech," yet we should ask ourselves whether, if it does occur in animals, any of our available methods of investigation would suffice to disclose it. *Prevarication* is one more criterion commonly advanced to set our species apart from other animals. This criterion will be discussed later in connection with the importance of intention in animal communication behavior.

Thorpe accepts the available ethological evidence, especially the recent studies of chimpanzees by the Gardners and by Premack, as convincing evidence that apes, at least, and probably also dogs and wolves, clearly demonstrate conscious purposiveness. He feels it is likely that, if the chimpanzee larynx were adequate, these apes could learn to speak as well as children three years old, or perhaps older. To Thorpe, "human speech is unique only in the way it combines and extends attributes which, in themselves, are not peculiar to man but are found also in more than one group of animals. . . . Yet . . . there comes a point where 'more' creates a 'difference'." Here he aptly quotes A. N. Whitehead (1938): "The distinction between men and animals is in one sense only a difference in degree. But the extent of the degree makes all the difference. The Rubicon has been crossed."

It is only fair to point out that many of the opinions discussed above date from the "pre-Washoe" period of ethology, and might not reflect the considered views which these authors now hold. Yet there is no doubt that for centuries philosophers and linguists have based their most fundamental definitions of humanity on very positive assertions about what animals can and cannot do. This means that whatever students of animal communication have learned, or can learn in the future, about communication behavior is directly relevant to major questions of fundamental significance to linguistics and philosophy.

Are Animals Aware
of What They Are Doing?

INSOFAR AS LINGUISTS AND PHILOSOPHERS have been correct in linking human thinking so closely to language, the communication behavior of other species is bound to suggest conscious thought to roughly the extent that it shares essential features with human speech. But behavioral scientists have been reluctant to consider this idea—primarily, I believe, because for half a century our thinking about animal behavior has been dominated by the behavioristic tradition in psychology (Lashley, 1923; Watson, 1929; reviewed by Lorenz, 1958, by Klopfer and Hailman, 1967, by Klein, 1970, by Stenhouse, 1973, and by Schultz, 1975).

Watson defined behaviorism operationally: "Behaviorism . . . holds that the subject matter of human psychology *is the behavior of the human being.* Behaviorism claims that consciousness is neither a definite nor a usable concept. . . . Its closest scientific companion is physiology. . . . It is different from physiology only in the grouping of its problems, not in fundamental or in central viewpoint." Biologists concerned with animal behavior have adhered, with very few exceptions, to the comparable tenets of Jacques Loeb (1900, 1912, 1916) and C. Lloyd Morgan (1894). Loeb was confident that virtually all behavior of animals and men could be analyzed successfully in terms of tropisms, or forced movements. Morgan's canon ("In no case may we interpret an action as the outcome of the exercise of a higher psychical faculty, if it can be interpreted as the outcome of the exercise of one which stands lower in the psychological scale") has been widely interpreted as requiring that complex functions should not be postulated if a simpler explanation will suffice. This is, of course, the widely accepted principle of parsimony; given a choice of two or more plausible explanations, the simplest is preferred.

Similar reasons of parsimony are usually given against any suggestion that animals might have mental experiences, or be aware of the flow of events with which their behavior interacts. As a result, only the most physiological explanations have customarily been recognized as worthy of scientific consideration. The behavioristic viewpoint will be further discussed and analyzed in Chapter 5, but it will be helpful first to review some general ideas about these challenging issues.

OPINIONS ABOUT AWARENESS IN ANIMALS

In allowing ourselves to entertain the notion that animals may be aware of past, present, and future events by means of mental images, in the sense loosely defined in Chapter 1, it certainly is not necessary to assume that such mental experiences are at all similar to those which a person might have under analogous circumstances. This uncertainty has been one of the chief reasons why many behaviorists have denied that it is meaningful to inquire about the properties of human mental experiences, let alone any that animals might have. But ethology has now developed to a point at which these behavioristic inhibitions may well have become a barrier to further progress, because they discourage recognition of significant opportunities for new and promising lines of investigation (Popper, 1974, p. 273).

Almost all linguists, and most philosophers who have considered the question, have vacillated between denying to animals any significant mental experiences (for example, Langer, 1942, 1962°,

1967, and 1972°), and grudgingly admitting the likelihood of certain simple ones, while rejecting others. The latter are then held to be of crucial importance. These discussions are often eloquent, but they show signs of what ethologists call conflict behavior. To some, this will appear to reflect a fundamental difference between scientists and humanists, but I am more optimistic, and suggest that communication behavior presents a magnificent opportunity for fruitful interaction and cross-fertilization between broad-minded scientists and equally perceptive humanists. The implicit denial of mental experiences to animals has almost become an act of faith, supported by arguments and assertions that true language is a unique and characteristic attribute of our species.

Brown (1958) opens a discussion of the comparative psychology of linguistic reference with a light-hearted but pointed paraphrase of current opinion: "I grant a mind to every human being, to each a full stock of feelings, thoughts, motives, and meanings. I hope they grant as much to me. How much of this mentality that we allow one another ought we to allow the monkey, the sparrow, the goldfish, the ant? Hadn't we better reserve something for ourselves alone, perhaps consciousness or self-consciousness, possibly linguistic reference?

"Most people are determined to hold the line against animals. Grant them the ability to make linguistic reference and they will be putting in a claim for minds and souls. The whole phyletic scale will come trooping into Heaven demanding immortality for every tadpole and hippopotamus. Better be firm now and make it clear that man alone can use language and make reference. There is a qualitative difference of mentality separating us from the animals."

Later (pages 164–171), Brown recognizes that "If vocalization is acknowledged to be unimportant, the dances of the bee appear to be very much like referential language." But he places great emphasis on the assumption that "the dances are unlike language in that they are not learned." He feels that animal communication is rigid, always predictable when the circumstances are specified, whereas human speech is not. He says of the dancing bees "the followers' reaction is *too* reliable."

The impression of mechanical predictability of the responses of follower bees stems in part from the eloquent simplicity of von Frisch's descriptions of this behavior in his semipopular books and articles. More detailed reading of his technical papers, or actual observation of the bees themselves, show considerable variability. Many followers do other things instead of flying out to the place indicated by the dances; they may seem to ignore the dances altogether and turn to other activities. Even those that do leave the hive do not all reach the indicated goal. It is technically so difficult to observe individual bees known to have followed a given dance after they are flying in the open air, that we know almost nothing about their behavior between the time they leave the hive and the moment they arrive at the feeding place. As pointed out by J. L. Gould (1975a, 1976), bees newly recruited by dances often take much longer to reach the food than the time necessary for a direct flight. Contrary to Brown's opinion, and that of many others, there is quite enough variability in the behavioral responses of bees and other communicating animals to leave room for the assumption of spontaneity.

Examples of "philosophical conflict behavior" are evident when Kenny, Longuet-Higgins, Lucas, and Waddington (1972, 1973) discuss the problems of analyzing mental processes from their respective viewpoints as philosopher, physicist, theologian, and biologist. When they consider the possibility that nonhuman minds might exist, they reflect the current climate of opinion by devoting much more attention to computers than to animals. At one point, Washoe and other chimpanzees are denied true minds on the ground that they merely mimic the sign language of the deaf, but elsewhere because they have been taught this language by human trainers. However, animals are conceded to have subjective feelings and perceptions, that is, sense-consciousness. Waddington argues that "if consciousness were to be adopted as a criterion of mind, it would be a signally useless one, because the only way to tell whether any other thing is conscious is to ask it. And that you can only do to human beings; . . . the concept of consciousness is not applicable to anything but a language using animal." Yet some animals, such as dogs and cats, are recognized as capable of having intentions, and Longuet-Higgins admits that: "An organism which can have intentions I think is one which could be said to possess a mind [provided it has] . . . the ability to form a plan, and make a decision—to adopt the plan." From this divergence of opinions, the presence of mental images and an ability to provide introspective reports on self-awareness and intentions emerge as a criterion of mind.

Hampshire (1959) clearly expressed the opinion of many philosophers: "It would be senseless to attribute to an animal a memory that distinguished the order of events in the past, and it would be senseless to attribute to it an expectation of an order of events in the future. It does not have the concepts of order, or any concepts at all. An intention involves, among other things, a definite and expressible expectation of an order of events in the future, and is possible only in a being who is capable of at least the rudiments of conceptual thought." Yet chimpanzees and other animals have been demonstrated to have at least moderately complex concepts. One example is provided by the laboratory experiments of Rohles and Devine (1966, 1967) on the "middleness" concept. Chimpanzees were trained to select from a row of objects the one in the middle position, in the sense that equal numbers of objects were to its left and right. If the objects were arranged symmetrically, chimpanzees could select the middle one when as many as 17 were present. If irregularly and asymmetrically arranged—that is, with irregular spacings between adjacent objects—chimpanzees could solve the problem when up to 11 were presented. This is roughly equivalent to the ability of four- to six-year-old children to solve comparable problems. Rohles and Devine did not succeed in training rhesus monkeys to select middle objects in sets larger than five.

The extensive field studies of ethologists have also cast grave doubt on negative assertions of Hampshire and many others. For example, ethologists have found the term "intention movement" widely applicable to those postures and relatively slight movements of animals that convey reliable information about their probable future behavior to other animals (Daanje, 1951; Tinbergen, 1951). There is no doubt from the reactions of conspecifics that such information is indeed conveyed. But it has been the curious custoᵢ of most ethologists to stop short of interpreting an intention movement as evidence that the animal has a *conscious* intention, although Daanje did state that animals intend to perform particular behavior patterns. Since both conspecifics and human observers can predict the future behavior of an animal from its intention movements, it seems remarkably unparsimonious to assume that the animal executing the intention movement cannot anticipate the next steps in its own behavior.

These reductionist attitudes are very resistant to erosion, however, and even the broadly perceptive biological commentator Lewis Thomas (1974) writes: "A solitary ant, afield, cannot be considered to have much of anything on his mind; indeed, with only a few neurons strung together by fibers, he can't be imagined to have a mind at all, much less a thought. He is more like a ganglion on legs." Is the dancing honeybee a ganglion on wings?

Bennett (1964) argued with charming erudition that to qualify as rational creatures bees would have to exchange more abstract messages than real bees had been demonstrated to do. But Bennett's ideas about real bees were apparently based on von Frisch's first reports of his discoveries, and Bennett did not consider the implications of the later discovery that the dances are used to communicate about things other than food or to exchange information about hive sites. As discussed by J. L. Gould (1976), many of Bennett's criteria for rationality, which he asserts are not satisfied by the dancing bees, have lost their validity, either because recent observations have shown that bee dances do satisfy the criterion in question, or because they have never been looked for by observations or experiments that could reveal either their presence or absence.

Black (1968), McMullan (1969°), Robinson (1973°), and others have distinguished animal communication systems from human language on the ground that the former are rigid responses to external or internal stimuli, which can, at least in principle, be definitely specified, whereas human language is spontaneous and creative. As pointed out above in connection with the views of Brown (1958), animal responses to communication signals are, in fact, quite variable. In this connection, it is commonly stated that no animal can use its communication system to tell a lie. Of course, a lie requires intention to deceive, so that to judge whether variability in animal communication behavior is "noise" or prevarication requires knowledge of the animal's intentions. But conscious intention is a category of mental experience that is widely believed to be uniquely human. Maritain (1957) exemplifies this climate of opinion: "Animals possess a variety of means of communication but no genuine language . . . no animal knows the relation of signification or uses signs as involving and manifesting an awareness of this relation . . . in the last analysis . . . the relation of signification remains unknown to the bees. They use signs—and they do not know that there are signs. . . . The whole thing belongs to the realm of conditioned reflexes, whereas language pertains to the realm of the intellect, with its concepts and universal notions." Grice (1957) states the generally accepted psychological conditions of language more rigorously: "Perhaps we may sum up what is necessary for A to mean something by x as follows. A must intend to induce by x a belief in an audience, and he must also intend his utterance to be recognized as so intended." There is a serious circularity in these arguments: conscious intention is ruled out *a priori* and then its absence taken as evidence that animal communication is fundamentally different from human language.

It seems that the more difficult the question under consideration and the less adequate the available evidence, the more definite become the generally accepted assertions about the differences between human language and animal communication. But it is important to ask on what basis such definite assertions can be made about what bees and other animals do or do not know. Have we allowed nonscientific value-judgments to color our thinking about these questions?

For example, Adler (1967) concludes that the communication system of honeybees is "a purely instinctive performance on their part and does not represent, *even in the slightest degree*, the same kind of highly variable, acquired or learned, and deliberately or intentionally exercised linguistic performance that is to be found in human speech." He then goes on to argue that if it were to be established by some future investigations that animals differ from men only in degree and not radically in kind, we would then no longer have any moral basis for treating them differently from men, and, conversely, that this knowledge would destroy our moral basis for holding that all men have basic rights and an individual dignity that render it wrong to mistreat groups of men judged to be inferior for the benefit of supposedly superior groups.

Followed to its logical conclusion, this argument implies that the comparative investigation of communication behavior has more dangerous potential consequences than nuclear physics had in the 1930s, or the current fear that synthesis of certain new forms of DNA might produce uncontrolled pathogens (Berg, et al., 1974). Unless we wish to abandon our scientific faith that understanding fundamental processes will prove of value to our own species, I suggest taking Adler's philosophical arguments as one more reason why it is important for scientists to investigate, as fully and accurately as we can, the relationship between human and animal communication.

The narrowing gap between animal communication and human language calls into question not only the assumption of a qualitative dichotomy, but also the accompanying assumption that animals respond mechanically to external or internal stimulation, whereas human beings speak with conscious understanding and intent. Despite the deep reflection and thoughtful eloquence of philosophers, the nature, and even the existence, of mental phenomena have remained largely outside the scope of natural science. Occam's razor and Morgan's canon have been so seriously adhered to since the 1920s that behavioral scientists have grown highly uncomfortable at the very thought of mental states or subjective qualities in animals. When they intrude on our scientific discourse, many of us feel sheepish, and when we find ourselves using such words as fear, pain, pleasure, or the like we tend to shield our reductionist egos behind a respectability blanket of quotation marks.

There have been a few exceptions to the behavioristic tradition and the related stress on parsimonious explanations in biology. Adams (1928) discussed what he felt were logical weaknesses in Morgan's original statement of his widely cited canon, which he found "insufficient to criticize the inference of mind." One problem with Morgan's canon is that it is based on an intuitive classification of behavior into higher and lower categories, the latter to be preferred unless the evidence forces postulation of the former. But no definitive, objective reasons are provided for assigning behavior to particular places on the scale of lower to higher. Adams advocated that mental experiences could reasonably be inferred in another animal to the degree that its "structure, situation, history, and behavior" resemble those which accompany such mental experiences in a human observer.

Jennings (1906, 1910, 1933) repeatedly and eloquently argued the case for an open mind concerning behavioral and mental continuity between men and other animals. Marler (1974) believes that "we delude ourselves if we think that a complete [behavioral] discontinuity separates us from other animals." And Boyle (1971) advocates that "the psychologist . . . will, if he is wise, acknowledge that organisms make sense of their experiences, and he must attempt to discover what this sense is. Unfortunately psychology has turned its back on this task because, since a psychologist's hypotheses about an animal's experience cannot be confirmed, psychologists have to a large extent ceased trying to understand other human beings as well." Yet Boyle, although ready to admit that when a cat rubs against his feet she intends to induce him to give her food, nevertheless finds it "difficult to imagine that a bee *intends* to communicate to fellow workers a message about the distance and direction of pollen. . . . That may be the meaning of the dance to an observer, but it is doubtful whether bees are capable of this type of understanding." Small size or phylogenetic remoteness from man are evidently taken as evidence against any form of conscious intent in bees. The image of a "ganglion on legs" dominates our view of invertebrate animals.

Tolman (1932) developed a "purposive behaviorism," in which what are ordinarily considered as mental events and processes were treated as intervening variables between external stimuli or internal influences on the one hand and overt, observable behavior on the other. By simply designating mental processes as intervening variables, the difficult question of their nature and reality is avoided. But Tolman's position is less rigid than strict behaviorism, and he accepts the reality of conscious awareness in animals, for example in a white rat at the moment of learning some new behavior pattern, such as a specific portion of a maze. In recent decades, however, most psychologists have either avoided this question or taken positions closer to strict behaviorism.

Many schools of philosophy dissent vigorously from materialism and from logical positivism, and not only do these schools accept the reality of concepts which behaviorists reject as meaningless; they often attach central importance to them. But philosophers of this kind seldom pay much attention to animals. There is an active discipline of cognitive psychology which feels free to deal with *human* mental experiences, although often refraining from the use of explicit mentalistic terms (Mowrer, 1960a, 1960b; Taylor, 1962°; Fodor, 1968; Estes, 1975). Neisser (1967) asserts emphatically that "Cognitive processes surely exist, so that it can hardly be unscientific to study them." Schultz (1975) reviews the trend for some contemporary psychologists to abandon the strict taboos of behaviorism. When Irwin (1974) considers many of these broad and challenging questions within a fairly conservative framework of objective analysis, he does recognize that men and animals have "expectancies." By this term, Irwin seems to mean something closely resembling mental or internal images of the possible future outcomes of various alternative patterns of behavior. The term consciousness, however, is still one which Irwin struggles to avoid, but Kimble and Perlmuter (1970°) go so far as to speak of volition. Fodor, Bever, and Garrett (1974) review the contributions of psycholinguists who have followed the lead of Chomsky in rejecting a behavioristic position regarding *human* language, and find it essential to consider mental entities. But few psychologists or even ethologists have yet moved away from an essentially behavioristic position with regard to animal behavior.

Lorenz (1958, 1963) is a notable exception among ethologists; he does not hesitate to express a belief that animals have subjective experiences, although, like Adams, he concentrates his attention on higher vertebrates. Bertrand (1969) hesitantly ventures to speak of a monkey's behavior as voluntary. Razran (1971) briefly considers the possibility that animals have simple thoughts. Brewer (1974) has recently argued in the course of a vigorous dissent from behaviorism: "Since cognitive theory holds for humans, it is unparsimonious not to apply it to animals." And Weiner (1972°) states that "cognitivists also use the man-animal continuity to promote their view that even infrahuman behavior is guided by cognitive processes." Mowrer (1960a, 1960b) struggled to escape the rigid restrictions of the behavioristic position while still dealing only with observable events, but was nevertheless led to state that "if consciousness were not itself experienced, we would have to invent some such equivalent construct to take its place." Yet he concludes that bee dances are "limited to 'sentences' of the thing-sign variety" because samples of food are transferred, rather than being represented by a symbol.

Wittgenstein (1953) approached these problems with Socratic questions, such as: "We say a dog is afraid his master will beat him; but not, he is afraid his master will beat him to-morrow. Why not?

"One can imagine an animal angry, frightened, unhappy, happy, startled. But hopeful? And why not?

"A dog believes his master is at the door. But can he also believe his master will come the day after to-morrow?—And *what* can he not do here? Can only those hope who can talk? Only those who have mastered the use of a language?"

The essence of Wittgenstein's skepticism seems to concern the time span of an animal's anticipation into the future, and he may well be correct in this estimate for a dog in the situation suggested. But suppose a dog *did* anticipate events likely to occur tomorrow. How could we recognize this fact from observing its behavior today? Lacking any evidence at all, we make a negative judgment, reasonably enough, but tend to forget its weak foundation.

Miller (1967) asserted that "Man is the only animal to have a combinatorily productive language . . . a species-specific form of behavior. . . . Serious attempts have been made to teach animals to speak. . . . These attempts have uniformly failed in the past and, if the argument here is correct, they will always fail in the future." Recent successes in teaching chimpanzees to communicate with combinations of gestures clearly cast doubt on the validity of Miller's "pre-Washoe" prediction. Rensch (1971°) and Popper (1972°) expressed similar views. Miller, Galanter, and Pribram (1960), Langer (1962, 1972), and others state without qualification that man is the only animal that can be aware of his own future death. But I suggest that we pause and ask just how does anyone know this? What sort of evidence is available either pro or con? Suggestive inferences can be based on the clear demonstration that many social animals recognize each other as individuals, and on the observation that some animal mothers show signs of distress over the corpses of their dead infants, which they carry about for days (Goodall, 1968, 1971, 1975). How can we judge whether an animal may experience any notion of its own future death after observing the death of companions (Cowgill, 1972)? The available, negative evidence supports at most an agnostic position.

EVIDENCE FROM NEUROPHYSIOLOGY

Eccles (1974) has proposed an explanatory framework for brain function that includes conscious experience, which he feels exists only in the human dominant cerebral hemisphere. He cautiously reserves judgment concerning the existence of consciousness in the subordinate hemisphere of the human brain and in the Great Apes. Sperry (1969) goes farther, and not only recognizes the importance of conscious experience in the human brain, but concludes that it "exerts a directive holistic form of control over the flow pattern of cerebral excitation." Much of the evidence on which Eccles and Sperry have based these ideas comes from surgical cutting of the corpus callosum in human patients suffering from severe epilepsy. When the hemispheres are thus deprived of their normal channel for exchanging information, they can learn to recognize quite separate sensory stimuli, and they seem to operate almost independently with respect to complex behavior. This is the case in both human patients and all other mammals tested so far. In some patients, only the dominant hemisphere can report verbally about its sensations and learned behavior. But in others, mental functions seem more evenly divided between the hemispheres (Nebes and Sperry, 1971; Teng and Sperry, 1973).

In some of the patients with a severed corpus callosum, the subordinate hemisphere can mediate learning to recognize objects, and the patient can demonstrate that learning has occurred by pointing to the correct object when asked to match a sample. Yet the same subject is unable to report the correct choice orally or in writing. It seems clear that the subordinate hemisphere carries out many of the mental functions ordinarily considered conscious, but lacks the ability to report them in words. As Eccles (1974) puts it, "the minor hemisphere resembles an animal brain." These discoveries add to the evidence for physiological continuity between men and animals in brain function, and suggest a comparable continuity in mental experiences. Pribram (1971) and Gazzaniga (1975) review much the same evidence and reach conclusions similar to those of Eccles, although they are expressed in slightly different terms.

DOES BEHAVIORAL COMPLEXITY IMPLY CONSCIOUS AWARENESS?

Many behavioral scientists express feelings of discomfort, or even outrage, at the inference of conscious intention in animals because previously unsuspected complexities in their orientation and communication have been discovered. On strictly logical grounds, complexity of behavior and conscious awareness are neither commensurate with or necessarily related to one another in any way. Inanimate mechanisms can be enormously complex and difficult to understand, but most *descriptions* of animal behavior can be modeled by mechanisms far simpler than a television receiver. The same can be said of many physiological mechanisms. For instance, a very simple electronic circuit can produce an electrical signal closely resembling the spike potential of a neuron. But only the most naive engineer-turned-neurophysiologist would accept the existence of such a circuit as a satisfactory explanation of the functioning of nervous systems. To Loeb, the existence of a phototactic machine constructed out of wheels, electric motors, and photocells was evidence for believing that animal, and even human, behavior could be explained in terms of tropisms. But in the 1970s, the crippling limitations of such intellectual mypoia should be clearly apparent; the simplicity often lies not in the behavior, but in its description.

Can we accept the reality of our own conscious awareness but reject the hypothesis that any pattern-recognizing machine is also consciously aware of the pattern to which it responds selectively?

It certainly seems easier to reject the notion of consciousness in a simple mechanism, such as a lock built to accept a particular key, than in a computer system programed to respond correctly to an especially intricate input signal. This is primarily because we can understand how the lock works, but find it impossible to encompass in our own mental imagery the complexities of the computer program. Must we therefore infer conscious awareness whenever we do not know how a mechanism works? Complexity of some kinds surely provides no convincing evidence for the existence of mental experiences. The physiological mechanisms by which a kidney regulates the chemical composition of the blood, or the biochemical systems that regulate cellular respiration, are marvelously complex and incompletely understood, but certainly are far different from central nervous systems in their structure and function. Analogies can be drawn by describing all three in terms so general that they apply to any self-regulating system, and these system properties are of interest in their own right. But brains and minds, insofar as we allow ourselves to admit the existence of the latter, are surely different from kidneys and mitochondria in basically significant attributes (Sperry, 1973°).

In the interpretation of communication behavior, pride of parsimony can lead us into some awkward situations. As mentioned briefly in Chapter 2, Smith (1975, and in press) has argued that the waggle dances of honeybees convey to other bees not information about distances and direction, or about the actual location of a distant food source, but only information about the internal state of the dancer; that is: "A forager bee does not state that it has found food at a certain place; rather, she describes a direction of flight (perhaps a flight she is likely to make again shortly) and, on request, provides a sample of the food." But suppose we knew nothing about human language and watched and listened to human beings while they were conversing. We might well conclude that they, too, were simply describing their internal states. Indeed, the definition of internal states can easily be extended to cover the most complex speech or writing, which can, if one wishes, be interpreted as "merely describing the internal state" of speaker or writer. Pushed to their limits, all these arguments make sense only on the implicit assumption of conscious intent on the part of human beings and its absence in animals.

The general feeling that our species is uniquely superior has suffered a series of intellectual setbacks beginning with the Copernican and Darwinian revolutions, the second having far more basic repercussions than the first. Later, the ability to learn from past experience was advanced as a unique human attribute, but successive discoveries of learning in animals more and more distantly related to us forced abandonment of that criterion. Tool-using suffered a similar fate, for example from studies of Darwin's finches and sea otters, as did tool-making more recently at the hands of chimpanzees (Goodall, 1968; McGrew, 1974). When evidence is presented that a nonhuman species achieves some of the previously proposed criteria for distinguishing human language, "the list grows longer in order to exclude the interloper species. If this kind of progression continues we may eventually have a definition of language that isomorphically maps the behavior of human language; in essence a redundant description of the behavior" (Fouts, 1973). Similar viewpoints have been presented in semipopular fashion by Linden (1974) and S. J. Gould (1975a).

Most biologists concerned with animal behavior seem to share the basic views of the linguists and philosophers quoted above—for example, Simpson (1964) and Dobzhansky (1967). When ethologists discuss animal communication, we almost never find any suggestion that subjective experiences may occur in the animals under study. The reductionist, behaviorist tradition still dominates our thinking. When von Frisch or Lindauer find that bees dance in patterns correlated with the location of food or something else needed by the colony, or that a bee which has danced about one potential location for the swarm shifts to a different site under the influence of the more enthusiastic dances of others, they do not suggest that perhaps the bees feel any need for sugar, pollen, water, or a suitable new cavity for the colony. That is taboo (Lindauer, 1955°, pp. 312–313). Virtually all descriptions and discussions are in terms that would be equally applicable to a living animal or an appropriately contrived machine.

This viewpoint has served our science well for more than 50 years by constraining speculations and focusing attention on phenomena amenable to experimental analysis. Seventy years ago, this disciplined restraint was a healthy reaction to an earlier tendency to ascribe human feelings to a wide variety of animals, mostly on the basis of rather unconvincing anecdotal evidence. But suppose the biologists and psychologists active at the turn of the century had known about the communication dances of bees and about the recent insights the Gardners have obtained from Washoe. Would they have been so adamant in banishing from respectable consideration all notions of mental states in animals? Their message was: As a working strategy of research, assume no mental states or subjective experiences, and see how much of animal behavior can be accounted for on this parsimonious basis. This has now been done on a large scale, and some of the results suggest that it is time to review our perspectives and strategies in the light of the new discoveries.

Is There Evolutionary Continuity of Mental Experiences?

The importance of this question is demonstrated by the heavy reliance of linguists and philosophers on the consciously intentional use of language as the principal distinguishing characteristic of our species. A major reason for these philosophical assertions has been the acceptance by those linguists and philosophers of the general conclusions expressed by students of animal behavior. I suggest that behavioral scientists now have the opportunity, and perhaps an obligation, to explore and discuss the implications for this traditional, behavioristic viewpoint of recent discoveries about communication behavior in animals.

When the question of subjective feelings or conscious intent in animals is raised, the customary response of many biologists or psychologists is to take the cautiously agnostic position that subjective or mental states, together with consciousness and intention, are beyond the reach of objective scientific inquiry. The virtual extinction of introspective psychology, under the onslaughts of behaviorism in the 1920s, seemed to be a firmly closed story. Yet, when the behavioristic position is stated at its scholarly best—for example, by Lashley (1923, 1958)—it is essentially agnostic. It does not deny the existence of mental states, but argues that they are one and the same as neurophysiological processes, and that it is unprofitable to attempt any sort of scientific analysis based on introspective reports. Half a century of behavioral science has progressed on this basis, along with many discoveries in neurobiology in the broadest sense, including ethology.

There remains that tendency for what was originally an agnostic position to drift implicitly into a sort of *de facto* denial that mental states or consciousness exist outside our own species. It is very easy for scientists to slip into the passive assumption that phenomena with which their customary methods cannot deal effectively are unimportant or even nonexistent. To quote Fouts (1973): "All one needs to do is to look around and *not* see something and then conclude that the thing that was not seen in a particular species is totally absent in that species." Here I should also like to follow the example of Holloway (1974) in quoting Daniel Yankelovich ("Smith," 1972): "The first step is to measure whatever can be easily measured. This is okay as far as it goes. The second step is to disregard that which can't be measured or give it an arbitrary quantitative value. This is artificial and misleading. The third step is to presume that what can't be measured easily isn't very important. This is blindness. The fourth step is to say what can't be easily measured really doesn't exist. This is suicide."

Biological evolution is universally accepted by behavioral scientists as a historical fact. Animals are used as surrogates or "models" for behavioral investigations on the implicit assumption that principles discovered in this way are applicable to our own species, as has been abundantly confirmed in physiology, biochemistry, and medicine. Certainly this assumption implies qualitative continuity. If, for example, human learning were believed to be radically different in kind from that available for analysis in other animals, no one would even suggest applying to questions of human education what has been learned by studying rats, pigeons, or monkeys. Yet, when questions of communication and language arise, even hard-nosed behaviorists take for granted a large element of discontinuity. To argue that language is unique to man and, therefore, no matter how complex animal communication turns out to be, it cannot possibly be continuous with human language, is indefensibly circular.

Must we reject evolutionary continuity in order to preserve our gut feeling of human superiority, as proposed by Adler (1967), Critchley (1960), and Langer (1967, 1972)? Or can we be satisfied with a merely quantitative, if enormous, difference between communication behavior in our own and other species? If we insist on a qualitative human-animal distinction in the area of communication behavior, a radical difference in kind in Adler's terms, must we support our insistence by criteria as subjective and difficult to test as those that were rejected by the founders of behaviorism?

The rigid position of the strict behaviorists is now being questioned with increasing frequency. For instance, Mower (1960b) introduces a chapter entitled "Images, Memory, and Attention (Observing Reactions)" with the remark that these terms have been "and perhaps are still, in some measure *taboo*. Many of us were taught, under pain of banishment from professional psychology, never to use these terms, at least not during 'working hours' . . . such language was deemed completely unsuited to the purposes of science. . . . But it is slightly ironical that those very methods of analysis and research which radical Behaviorism introduced are now leading us, ineluctably, back to concepts which Behaviorism was determined to ignore—or even destroy."

In facing squarely the problems of dealing with the possibility that animals have mental experiences, it may be helpful to recognize that our current climate of opinion in the behavioral sciences involves a gradient of acceptability concerning the terms and concepts listed below:

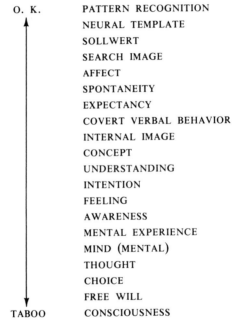

O. K.	PATTERN RECOGNITION
	NEURAL TEMPLATE
	SOLLWERT
	SEARCH IMAGE
	AFFECT
	SPONTANEITY
	EXPECTANCY
	COVERT VERBAL BEHAVIOR
	INTERNAL IMAGE
	CONCEPT
	UNDERSTANDING
	INTENTION
	FEELING
	AWARENESS
	MENTAL EXPERIENCE
	MIND (MENTAL)
	THOUGHT
	CHOICE
	FREE WILL
TABOO	CONSCIOUSNESS

Individual scientists might wish to rearrange some entries in this rank order of orthodoxy, but there is no doubt that the gradient is a significant reflection of the current Zeitgeist. Rearranging these terms like playing cards is an entertaining game, but few radical rearrangements would leave the list a plausible one. It is also instructive to ask where one should draw a line to represent the boundary of scientific validity. Very strict behaviorists might stop after Affect, others may venture farther down the list. There are, of course, many philosophers who disagree with positivism, and they feel comfortable with a list extending beyond this one in the direction labeled Taboo (Fodor, 1968°; Fiegl, Sellars, and Lehrer, 1972°; Polten, 1973°).

Perhaps Jennings and Thorpe have outlined the most reasonable view, considering the limited evidence available; namely, that the gradient is a true continuum without sharp discontinuities. Furthermore, it seems more likely than not that certain animals have mental experiences involving, to varying degrees, the attributes represented crudely by this rank-ordered list of terms.

Many branches of science have made significant and substantial progress by employing postulated entities that could not be observed directly, at least when first developed, but which were inferred from observations of their supposed causes and effects. Gravitation, electric potentials, magnetic fields, atoms, neutrinos, X-rays, chemical bonds, hormones, genes, nerve impulses, and magnetic monopoles are pertinent examples. The impossibility of neatly verifying the existence of mesons or quarks has not inhibited theoretical physicists. Astrophysics is likewise based on concepts about events and processes immune from direct observation by any methods we can yet imagine.

Investigators of behavior have attempted to formulate comparable explanatory concepts, such as motivation, drives, or Lorenz's specific-action potential. But perhaps we have been overlooking more directly pertinent concepts lying close to hand or even closer—inside our own heads. When thinking about Washoe in the act of exchanging information about objects, actions, or desires via coded manual gestures, or when contemplating Lindauer's swarming bees dancing about the suitability and location of cavities where the swarm might find a new home, I submit that it may actually clarify our thinking to entertain such thoughts as "Washoe *hopes* to go out for a romp, and *intends* to influence her human companions to that end," or "This bee *likes* one cavity better than the other, and *wants* her swarm to occupy the preferred one."

Of course, the use of such terms as *want* or *like* does not explain the basic causes of the observed behavior or of any mental experiences that may accompany it. Nor should the use of these or similar terms be taken to imply identity with any human mental experiences. The degree of similarity or difference would be an appropriate question for future investigations. Perhaps this return to a consideration of basic subjective qualities can supply a unifying framework into which many complexities of animal behavior can be fitted. To paraphrase Boyle (1971), perhaps we can understand how, and to what extent, animals make sense of the flow of events of which their behavior forms a part.

Most people not indoctrinated in the behaviorist tradition take it for granted that animals do have sensations, feelings, and intentions. This intuitive impression is based on our experience with patterns of animal behavior that appear sufficiently analogous to some of our own behavior to permit us to emphathize. The dilemma of contemporary behavioral scientists results from our indoctrination that *as scientists* we must put such notions behind us as childish sentimentality unworthy of a rigorous investigator (Hebb, 1974). Yet the behavioristic and reductionistic parsimony typified by Watson and Loeb may have led us down a sort of blind alley, at the end of which we find ourselves defending to the last, at least by implication, a denial of mental experience to animals, a denial which we cannot justify on any explicit basis except the presumed absence of communication with conscious intent. As learnedly discussed by Malcolm (1973°),even Descartes, the fountainhead of the philosophical view that animals are merely machines, admitted that they could feel pain or pleasure and express passions. Yet many behaviorists believe that it makes no difference whether one thinks in terms of possible mental experiences or simply in terms of stimuli and responses, however complex. The same argument can be applied to people.

Inasmuch as we have only indirect evidence about *their* mental experiences, we may logically question whether these really exist. But if questions are raised by others about the reality of one's own subjective feelings, who is likely to fall back on a negative, or even an agnostic, response?

THE NATURE AND NURTURE OF MENTAL EXPERIENCES

One consequence of the view that awareness is simply one aspect of neurophysiological processes is to raise the nature-nuture question with regard to mental experiences themselves. To the extent that they are dealt with at all by scientists, it seems to be tacitly assumed that mental experiences result solely from individual experience and, in particular, from learning. This implication is clear in the statements of Pollio (1974), Maritain (1957), and Adler (1967), quoted above.

If we really accept the behaviorists' axiom of psychoneural identity (discussed from several viewpoints in the volume edited by Feyerabend and Maxwell, 1966), we should face up to the possibility that a nervous system might attain those properties leading to mental experiences primarily on the basis of genetic information. The development of mental experiences might depend on environmental influences only in the general and unspecific sense that DNA cannot lead to a complete animal without an environment that provides the necessary nourishment and other conditions. We might therefore conclude that the behavioristic argument leads to the likelihood of something akin to "innate ideas" in the philosophical sense.

The nature-nuture issue with respect to behavior has aroused some of the most violent passions and heated debates known among scientists. As is usual in such cases, balanced consideration strongly suggests that both sides are partly correct and that individual experience and genetic heritage each has significant effects on behavior (reviewed by Brown, 1975). It is also obvious that the relative importance of these two major factors varies widely between behavior patterns and groups of animals. This means that two extreme, all-or-nothing positions are clearly and equally untenable. Furthermore, interactions between genetic and environmental factors are of considerable importance, and this, together with the enormous difficulty of controlled experiments, make it impracticable in most cases to estimate their relative importance with anything approaching adequate accuracy. This does not mean that either can safely be assumed *a priori* to be all-important (or unimportant) in the absence of direct evidence of a sort that is rarely available at present.

Applying the same balanced approach to mental experience leads to a cautiously open mind concerning the possibility that both genetic and environmental influences, and interactions between them, may be important in the causation of mental processes, including awareness. Knowing so little about mental experiences in other species, we can scarely begin to attack the nature-nuture question, despite its potential importance. But we should not overlook the reality of the question any more than it seems sensible to ignore the possible existence of mental expressions in more than one species.

5

Objections
and their Limitations

To reopen the questions discussed in this book that have long seemed irrelevant to twentieth-century behavioral science is disturbing in many ways. Many would clearly prefer to pack all of these notions back into the secure Pandora's Box where they have quietly rested for so many years. Others appear simply to prefer statements of faith that man is radically different in kind from all other animals and, furthermore, is intrinsically superior, not only mentally but in fundamental moral values. These deep-seated objections deserve careful attention, for it is surely no accident that they are so widely and strongly felt.

The Behavioristic Objection

Virtually all comparative psychologists—and ethologists, as well—are at least *de facto* behaviorists in the sense that they concern themselves only with observable behavior and shun any involvement with possible subjective qualities or mental experiences (Lashley, 1949°). The strict behaviorist believes that it is operationally meaningless, and hence foolishly unscientific, to consider even human mental experiences, because they cannot be observed directly and because verbal reports about them are inconsistent and impossible to verify by any other means. S. R. Brown (1972) has clearly stated this dilemma. "Human subjectivity is a phenomenon that has both interested and eluded social scientists for some time. . . . Perhaps beginning with Watson, the behaviorists rejected introspection and romantic mentalisms, and they did so for sound scientific reasons: As scientists, they knew no way of dealing with these mental goings-on. The rejection, however, was regarded at the time (except, perhaps, in the case of Watson himself) as temporary, pending the development of instrumentation capable of dealing with what previously had proved elusive. The second generation behaviorist forgot the original reasons for the rejections, remembering only the act of rejection itself; the third generation behaviorist was even at a greater disadvantage, being unaware anything was even forgotten."

When strict behaviorists asserted that "The supposedly unique facts of consciousness do not exist" (Lashley, 1923), they were rebelling against schools of psychology which held that mental qualities were different in kind from physiological processes and hence could never, in principle, be explained in physicochemical terms. Lashley was attacking what he saw as the subjectivists' belief in a separate psychic world, "a unique mode of existence not definable in objective terms." He viewed introspection as "an example of the pathology of scientific method." But he also conceded that "there can be no valid objection by the behaviorist to the introspective method so long as no claim is made that the method reveals something beside bodily activity."

Hebb (1974) has recently argued vehemently: "Subjective science? There isn't such a thing. Introspectionism is a dead duck." Yet later in the same article he asserts that "Psychology is about the mind: the central issue, the great mystery, the toughest problem of all." But it is not clear how a central problem can be solved by ignoring most or all of the available evidence. B. F. Skinner (1957) was well aware of the danger that important problems might be ignored merely because they are difficult to study. He attempted to deal with what are generally called mental processes, or thinking, while maintaining a consistent behavioristic position. This he did by treating thinking as *covert verbal behavior*, in which the speaker and listener are the same person. This position equates thinking with a sort of talking to oneself, which is almost as unobservable as the mental concepts which Watson and the original behaviorists rejected as unworthy of scientific mention. But Skinner sees an important difference, in that such covert verbal behavior can be influenced by prior and subsequent events, especially by reinforcement, and thus, like other behavior, its properties can be deduced from an analysis of its antecedent causes and subsequent results. Because Skinner's behaviorism is constrained to deal only with input-output relationships and methods of reinforcement, it was possible, by stretching its definitions, to include even unobservable, covert verbalizing; at least, Skinner felt justified in doing so.

For those who cannot bring themselves to consider the possibility that animals can have even the simplest of thoughts, it may be more acceptable to follow Skinner's lead, despite the searching criticism and rejection of a behavioristic treatment of human language by Chomsky (1959). For instance, Skinner's definition of thinking as covert verbal behavior can reasonably be extended to include not only the literal exchange of words and sentences, but also the many other forms of communication used by human beings, such as signs, gestures, mathematical symbols, and the whole nexus of nonverbal communication. Lashley (1923, p. 342) anticipated important elements of Skinner's position, including this broadening of its scope: "The relation of any integration to the speech and gestural mechanisms is of prime importance for its 'conscious aspects.' Not only is the single certain evidence of consciousness in another person the existence of consistent, rational expressive movements. . . . The core of 'conscious' integration is the verbo-gestural coordination." We can now include consistently communicating animals within Lashley's definition, on the basis of new discoveries. Extensions of this approach are summarized in McGuigan and Schoonover (1973).

Including nonverbal as well as verbal communication in a behavioristic definition of thinking, along the lines advocated by Lashley and Skinner, seems intuitively to be even less of an extension of the behaviorist position than the redefinition of thinking as covert exchange of words and sentences. To accept into the covert fold verbal, but not nonverbal, communication would scarcely be parsimonious. Extending Skinner's concept of covert verbal behavior to covert communication behavior internally within the human brain, we may then ask, why not also within the chimpanzee brain? Washoe and the other chimpanzees who have learned to use and combine gestures that serve them much as the words in human speech can be assumed to manipulate covertly whatever communication systems they use overtly. Can birds sing to themselves covertly, and bees dance to themselves when motivation for communication behavior is present but the overt behavior is not possible for some reason? Some readers may feel they have been led astray and that something must be wrong, since we have reached the unwelcome conclusion that animals can think. But where is the error? Is it in the argument that if words can be employed covertly, so can nonverbal means of communication? Or in carrying over the notion of covert communication from people to chimpanzees? Is this notion acceptable in apes, but not in birds or bees, and, if so, where and how is a line to be drawn?

One objection that can be anticipated is the assumption that animals always express their motivational states immediately, whereas men can inhibit the expression and yet retain the internal communication behavior. But this seems highly unlikely as a general rule, for in many cases animals clearly retain the memory of some relationship about which they communicate only when circumstances are appropriate. One example from honeybees may suffice to make this point. If bees from a colony with a severe need for food have been foraging at a certain food source and dancing actively about its location and quality right up to sundown, they will ordinarily do no dancing during the night, but as soon as morning comes they will fly out to the same source, taking now a very different direction relative to the sun (Frisch, 1967). One could explain this behavior by postulating, quite reasonably, that they had learned other cues to the location of the food, such as landmarks. But under some circumstances, such bees can be stimulated to dance during the night if appropriate lighting conditions are provided. They dance with the waggle runs oriented at angles to gravity intermediate between the directions indicated in the evening and morning, with the difference roughly proportional to the time that has elapsed since sunset. Not only does this demonstrate the existence of an endogenous biological clock and continuous correction for the passage of time; it also suggests that the memory of food location and the motivation to communicate about it remain present in a latent, covert state somewhere within the bee's nervous system. Covert communication behavior can thus be postulated and its properties deduced from the effects of prior causes and subsequent results, just as Skinner attempts to do with covert human verbal behavior considered as an objective definition of thinking.

The assumption that animals always respond in a rigid, machinelike fashion to immediate stimuli is widespread and resistant to the influence of extensive evidence that animals often defer responses to stimuli. This evidence comes both from laboratory experiments, in which animals can be trained to respond not immediately, but after varying intervals, and from ethological observations under natural conditions. In general, animals we consider "higher" can learn to make such responses after longer intervals of delay.

The views of Terwilliger (1968) exemplify how this tendency to think of animals as Cartesian machines is often based on questionable implicit assumptions. Arguing that language is a uniquely human trait, he dismisses the dances of honeybees as something inferior because of their assumed rigidity: "no bee was ever seen dancing about yesterday's honey, not to mention tomorrow's. . . . Moreover, bees never make mistakes in their dance." Leaving aside the technical error that the dances concern nectar, rather than honey, Terwilliger was either unaware of, or ignored, much of the evidence reviewed in Chapter 2, along with the fact that bees can be stimulated to dance during the middle of the night about a food source they have visited the day before and will almost certainly visit again the next morning. These questions are discussed in more detail by Gould (1976).

The views of Mowrer (1960b), quoted in the previous chapter, indicate that many psychologists are relaxing the rigid strictures of Watson's original behaviorism. An agnostic reservation of judgment is clearly the soundest position at present. But it should be an open-minded agnosticism, which recognizes the possibility that anwers not yet available may be obtained from future investigation of questions about mental experiences in animals.

THE ANTHROPOMORPHIC OBJECTION

Against the notion that animals have mental experiences, it is often objected that such thinking is anthropomorphic, that is, that it requires ascribing human thoughts to other species. But consider the alternative hypothesis—that mental experiences, like other attributes of animals and men, exhibit a continuity of variation and are not typologically discrete, all-or-nothing qualities totally restricted to a single species. There is no reason to believe that any mental experiences that animals may have must be identical to our own. Indeed, there is no reason to suppose that they are absolutely identical throughout *Homo sapiens*, for instance between men and women, or children and adults. As Wittgenstein (1953) puts it, "one human being can be a complete enigma to another. We learn this when we come into a strange country with entirely strange traditions; and, what is more, even given a mastery of the country's language. We do not *understand* the people. (And not because of not knowing what they are saying to themselves.) We cannot find our feet with them."

It is actually no more anthropomorphic, strictly speaking, to postulate mental experiences in another species than to compare its bony structure, nervous system, or antibodies with our own. Subjective qualities have remained largely untouched by the Darwinian revolution, primarily for lack of effective methods for detecting them reliably in other species, let alone analyzing them by scientific methods. The prevailing view implies that only our species can have any sort of conscious awareness or that, should animals have mental experiences, they must be identical with ours, since there can be no other kind. It is this conceit which is truly anthropomorphic, because it assumes a species monopoly of an important quality. The attitude resembles, in many ways, the pre-Copernican certainty that the earth must lie at the center of the universe.

When we attempt to imagine what an animal may be thinking in a given situation, we are obliged to describe in words (or perhaps in human gestures or expressions) what we suggest the animal may feel. Critics object that the mere use of human language introduces a crippling error, because the animal does not use words. But this may be no more than a methodological problem that could be overcome by careful procedures, as, for example, in recent experiments with chimpanzees. The objection would be critical if, and only if, human thoughts were absolutely identical with words and never had a form not conveyed precisely by a verbal description, and if no other kind of thought existed. It seems highly unlikely that both these statements are correct, even for our own species (Hutchinson, 1976). They imply, for example, that new experiences could never occur in the absence of appropriate words already shared by a group of people speaking the same language. This would be a serious limitation of the creativity held by Chomsky and many others to be such an important attribute of human language. Granting that words express thoughts more or less imperfectly, the residue of properties not absolutely coextensive with words might be supplied by covert nonverbal communication in the sense discussed earlier.

This possibility is more difficult to evaluate and, by extending the concept of covert nonverbal communication, one might well be able to include all vague feelings, impulses, emotions, and so forth. But whether this can be done in any convincing fashion is not critical to the present discussion. I suggest that it is more likely than not that thoughts or mental experiences in people and animals share important properties without being completely identical.

A further objection that may confidently be anticipated is that interspecific communication could occur only between very closely related species, say man and chimpanzee, but not man and dog and still less man and bird, while man-to-bee exchanges of information would be quite unthinkable because of the remoteness of any common ancestors. This widely held view contains the implicit assumption that mental experiences vary rapidly with branching evolutionary lines of descent. But the mere possibility that mental experiences might exist in animals has been so thoroughly ignored by behavioral scientists that we naturally have no data whatever about how such experiences may vary between species.

An equally plausible hypothesis is that, as mental experiences are directly linked to neurophysiological processes—or absolutely identical with them, according to the strict behaviorists—our best evidence by which to compare them across species stems from comparative neurophysiology. To the extent that basic properties of neurons, synapses, and neuroendocrine mechanisms are similar, we might expect to find comparably similar mental experiences. It is well known that basic neurophysiological functions are very similar indeed in all multicellular animals. On this basis, we might be justified in turning the original argument of the strict behaviorists completely upside down. Because neurophysiological mechanisms appear to be very similar in men and bees, the mental experiences resulting from their operation must, according to this line of reasoning, be equally similar. If this seems an embarrassing conclusion, we can try to escape from it by postulating that neurophysiology is seriously incomplete, having failed, so far, to locate those functions that differ so widely between taxonomic groups that they generate incomprehensibly divergent mental experiences. The alternative is to postulate some special form of human uniqueness, not demonstrable objectively—an unparsimonious procedure, to say the least.

THE DANGERS OF RELAXING CRITICAL STANDARDS

Opening our minds to consider the possibility that at least some animals may have mental experiences is only a first step, though a crucial one that requires a significant departure from the current *Zeitgeist* of the behavioral sciences. Having taken this step, one is tempted to plunge at once into the very different process of inferring particular mental experiences in specific animals. This second step is as hazardous as it is seductive. It is so easy to guess about the processes occurring in the brain of another species, on the basis of its observed behavior, that we can easily forget how many such confident conjectures have come to seem implausible as more has been learned about the natural behavior of the animals concerned. Careful ethological analysis of preceding and subsequent behavior can provide some indications of the possible mental experiences that may exist within a given animal under particular circumstances. But we would be on much firmer ground if we could obtain more direct access to whatever events may be taking place inside the animal's head. Some possible approaches to this difficult task are suggested in Chapter 7.

The reductionist and behavioristic reactions which have led to the inhibitions discussed above—inhibitions which it now seems timely to relax to some degree—arose for the excellent reason that unsupported and highly implausible interpretations of animal behavior had become widespread toward the end of the nineteenth century. George Miller (1962) has pointed out that Morgan's canon was proposed and achieved wide acceptance, not with any intention of avoiding introspection, nor with any doubt that mental experiences exist in both men and animals. Rather, as

Miller puts it, "all that Morgan hoped for were a few reasonable rules for playing the anthropomorphic game." The view that it is anthropomorphic to postulate any sort of mental experiences in animals may have resulted from a confusion of scientific caution and parsimony with unscientific feelings of human superiority. Nevertheless, there is no doubt that enthusiastic observers of animals are constantly in danger of interpreting their behavior in more complex terms than is necessary or correct. Clever Hans is an outstanding example. It seems inherently reasonable that the ability of a horse to notice when a human experimenter had stopped making small counting movements and was waiting expectantly for the horse to end its tapping, is a simpler and far more convincing interpretation than the conclusion that this or other horses could really carry out complex arithmetical manipulations, such as multiplication and division of numbers written on a blackboard.

Thus, an important dilemma must be faced if we allow ourselves, contrary to the exhortations of strict behaviorists, to postulate, however tentatively, the existence of mental experiences in animals. To recognize that any mental experiences animals may have need not be identical, or even necessarily similar, to those of a man under comparable conditions, opens up a wider range of potential interpretation but, at the same time, makes it more difficult to gather convincing data. In Chapter 7, I suggest possible solutions to this dilemma. For the time being, it must suffice to emphasize that the problem is serious. Cautious treading of a middle ground is clearly called for to avoid both of two obviously fallacious extremes: (1) the postulation of complex mental activites (such as horses capable of long-division) when simpler ones are consistent with the observed behavior of the animal and the observed responses of conspecifics to its communication signals; and (2) the conventional reductionist position that animals have no mental experiences at all, or that any they may have are hopelessly inaccessible to our investigation.

Admitting into our spectrum of hypotheses worthy of consideration the possibility that animals have mental experiences, involves a responsibility to be extraordinarily cautious before concluding that a given animal has a particular kind of mental experience under particular circumstances. As we begin to investigate the possibility that mental experiences play a significant role in animal behavior, we should be warned by the unjustified definiteness of the quotations cited in Chapters 2 and 3.

THE "SO WHAT?" OBJECTION

One vestige of strict behaviorism takes the form of asking disparagingly what difference it would make in our ideas about animal behavior, or our investigations of it, if we *did* postulate mental experiences in the animals under study. Surprisingly enough, many ethologists have taken over this aspect of behaviorism. To be sure, some investigations and conclusions will depend very little on the distinction between a Cartesian concept of animals as purely deterministic, unconscious machines, and a broader view, which accepts the possibility that conscious intentions and mental experiences may be present. One can derive predictive generalizations by considering only one aspect of the phenomena under study. For example, one can predict the caloric value of plant or animal tissues by burning them in bomb calorimeters, and such experiments have demonstrated the important fact that the energy released by oxidation of living tissue can be predicted by the same chemical principles that govern the heat of combustion of inorganic substances. It is completely irrelevant to such investigations whether the living tissues come from animals, plants, brains, leaves, roots, ovaries, or fingernails. If one is interested only in heat of combustion, one can ignore these distinctions.

This example is a reduction to the absurd, but when we are dealing with phenomena as complex and subtle as those of behavior and social communication, it is prudent to keep an open mind concerning which attributes of the systems under study may be important. If one is interested only in relatively straightforward predictions of simply described aspects of behavior, the Cartesian viewpoint may be wholly sufficient.

On the other hand, there is considerable reason to believe that in at least one species, *Homo sapiens*, mental experiences play a significant, though not all-encompassing, role in the regulation of behavior. Accepting the reality of our evolutionary relationship to other species of animals, it is unparsimonious to assume a rigid dichotomy of interpretation which insists that mental experiences have some effect on the behavior of one species of animal but none at all on any other. It would be absurd to deny that mental experiences are important components in human behavior and human affairs in general. To the extent that animals have them, mental experiences may also be significant in their activities. It is obvious that one could not understand human beings as well, or predict their behavior as accurately, without taking some account of their awareness and intentions. The same consideration applies to other species, insofar as mental experiences are significant in their behavior.

An Obsolete Strait Jacket

Ignoring the possible existence of mental experiences and conscious intent in animals may have held back our scientific progress in this important field, as anticipated by Jennings (1933). Consider, for example, von Frisch's discovery of the dance communication of honeybees described in Chapter 2. In the early 1920s, he noticed that round dances occurred when he offered sugar solution in dishes, whereas waggle dances were correlated with the gathering of pollen. His papers did not mention the possibility that a communication system might convey anything more complicated than the type of food being brought to the hive. Only some 20 years later did he discover the far more significant correlation of dance pattern with distance to a source of food (Frisch, 1946). It seems possible that the communication dances of bees might have been discovered in the 1920s, if anything like complex communication among insects had not been so utterly unthinkable. This question leads to another: What are we *now* overlooking, as a result of comparable restrictions imposed on the questions we ask, by our basic viewpoint about the nature of animal and human behavior?

The extensive data gathered by behavioral scientists often seem to be filled with confusing contradictions; perhaps unifying concepts might be more easily discerned if mental experiences were included within the scope of our hypotheses and explanations. If so, their recognition would indeed make a difference. The first example that comes to mind is the injury-feigning, or predator-distraction, displays of certain birds. Ordinarily, this type of behavior results when an adult bird incubating eggs or caring for its young is disturbed by the approach of a man or other potential predator. The bird acts as though it is injured, flutters in a manner suggesting that it has a broken wing, and tends to move slowly away from the nest or young. Often the potential predator does, in fact, follow the ostensibly injured bird and, after this process has been continued over an appreciable distance, the bird suddenly recovers its normal capabilities and flies away (Brown, 1962; Gramza, 1967; Wilson, 1975; Skutch, 1976).

During the past half-century, ornithologists and ethologists have gone to great lengths to deny that such birds could have any conscious intention to lead the potential predator away from its offspring. For example, in a general review of predator-distraction displays, Armstrong (1949) feels it is important "to have available a series of terms that are devoid of disputable cognitive implications." The term injury-simulation is preferred to injury-feigning, because it "carries the sense of deliberate intention to deceive rather less than 'feign' and is, therefore, preferable." Armstrong suggests that "distraction displays have arisen through the 'displacement' of components from other behavior contexts, particularly threat and epigamic display, which have become ritualized into new behaviour-patterns with survival value." This strongly implies that a ritualization of displays which originally served other purposes is an adequate explanation and, furthermore, one which removes any need to be concerned about whether the birds are consciously aware of the probable results of this distinctive display. It is quite possible that both interpretations are correct; that an animal can consciously employ ritualized display patterns which have the evolutionary origin suggested by Armstrong and others. No adequate data are available to settle these questions, but they are worth asking.

A second example in which the possible presence of mental images may help make sense out of behavior patterns that otherwise require complex explanations are the elaborate structures built by certain species of birds as part of their mating displays. Some of the most extreme examples are the bowers of the bower-birds and their relatives in Australia, New Guinea, and adjacent islands (Marshall, 1954). In some species, the bowers are remarkably complex structures built from twigs, grass, and other vegetation in small areas which the bird has cleared. They are often decorated with conspicuous objects, such as fruits, flowers, fungi, and occasionally with silver coins, jewelry, or even automobile keys. The males display at these bowers and females are attracted by the displays, which presumably play a significant role in mate selection. Marshall and others are vigorous in their denials of any interpretation of bower-building which implies conscious intent on the part of the bower-bird. "These and other singular attributes have caused a voluminous popular literature to spring up about the family. Much of this is nonsense. Most of it has been marred by anthropomorphic generalization, and all of it is unsupported by experimental evidence." Marshall is emphatic that his aim is "to describe these and associated phenomena in terms of animal rather than human behaviour. . . . These complex and remarkable phenomena are probably expressions of innate behaviour patterns that are annually called into play by the secretion of sex hormones. . . . The theories of Australian naturalists that bower-birds are especially intelligent and that their display activities are largely 'relaxative', consciously aesthetic, and unconnected with the sexual drive are rejected, though of course it is not suggested that the birds do not enjoy the fantastic activities that they perform." Hartshorne (1973) is likewise inclined to believe that, in song birds, "Song expresses feeling, according to principles partly common to the higher animals. That a bird sings 'because it is happy' is not entirely foolish."

In the final chapter of his book, Marshall returns to questions of intelligence and estheticism in bower-birds. "While I have ascribed a utilitarian basis for each of the behavioral phenomena discussed, I see no reason, provisionally, to deny that bower-birds possess an aesthetic sense although, it must be emphasized, we have as yet no concrete proof that such is the case. Some bower-birds certainly select for their displays objects that are beautiful to *us*. Further, they discard flowers when they fade, fruit when it decays, and feathers when they become bedraggled and discoloured. But, it must be remembered, however beautiful such articles may be, they are still probably selected compulsively in obedience to the birds' heredity and physiology." Marshall thus assumes that bower-birds behave "compulsively" and, furthermore, without conscious awareness of the results of their behavior.

Bower-bird bowers are extreme cases out of a huge spectrum of behavior in which animals alter their immediate environments by constructing shelters or arenas used for mating displays. We have become so accustomed to concentrating on functional and adaptive aspects of these behavior patterns that we have neglected even to ask whether the animals have any awareness of the probable consequences of their behavior. While we do not yet have available direct evidence indicating whether they do or do not intend to influence perceived future events, it may be a serious limitation in our thinking to assume *a priori* that no such awareness can possibly exist.

As pointed out in Chapter 4, the possibility that animals have subjective feelings of various kinds has been ignored as studiously as has the possibility that they might have intentions or mental images. Behavior such as building and decorating bowers or feigning injury raises questions about both the feelings and the mental images that might precede or accompany such behavior. As discussed by Lorenz (1963), the subjective feelings of animals, while even more difficult to study than any mental images they may have, could well be of equal or greater importance. Indeed, most scientists, like other thoughtful people, have little difficulty in accepting the notion that injured animals feel a sort of pain, or starved animals a kind of hunger, akin to the comparable human feelings. The detailed analysis of animal feelings is another important challenge for ethologists, but one that lies outside the scope of this book.

Perhaps animals perform some of the behavior patterns we observe because they enjoy the resulting experience. For instance, herring gulls often soar for hours back and forth over the same area where there is an obstruction updraft with no realistic prospects for food and no evident social function. Behavior such as soaring on obstruction updrafts may be adaptively neutral, or virtually so, but result in a pleasantly satisfying feeling on the bird's part. One can even postulate that pleasant feelings that result when a physiological capacity is exercised are in themselves adaptive. Even though we cannot yet formulate such concepts as pleasure at all adequately in terms of physiological or biochemical correlates, this does not seem to me to be a sufficient reason for avoiding the concepts themselves, as though they were a dangerous plague. Once one begins to think about the possibility that animals have mental experiences, a number of other behavior patterns come to mind which might be explained more readily, and even more parsimoniously, if mental experiences were used as parts of our theoretical explanations. The possible ramifications of this line of thought are too extensive for adequate discussion in this book, and their exploration in both theoretical and empirical terms will both require and warrant extensive and challenging investigations.

To state that animals do something because they enjoy it is often criticized as tautological. Such criticism argues that the postulation of enjoyment adds nothing to the simple statement that the animal performs the behavior pattern in question. The critic may also go on to object that stating that the animal enjoys a given activity brings one no closer to an ultimate explanation of why it does so. These criticisms are valid only if applied to the belief that by postulating enjoyment one has produced an ultimate explanation of the cause for the behavior or the mental experience. But to recognize the probable reality of some attribute is not at all the same as asserting that it constitutes a complete causal explanation. It seems likely that some deep-seated reluctance to think about animal awareness underlies this type of objection to considering, even in the most tentative fashion, that animals may have mental experiences.

I have raised questions that neither we nor our descendants may be able to answer in the foreseeable future. Yet scientific progress clearly requires the formulation of alternate hypotheses before the most appropriate questions can even be asked. The customary approach to animal behavior tends to rule out, in advance of any investigation, the possibility that a system of animal communication may be more complex and subtle than is demonstrated by the data immediately at hand, still less that it may involve conscious intention. If the communication system is found to be variable and conveys many fine distinctions, those can easily be dismissed as "noise" of no significance, because the thinking of the experimenter and the resulting nature of his experiments have, so to speak, too coarse a grain.

6

The Adaptive Value
of Conscious Awareness

THAT SOCIAL COMMUNICATION is adaptively valuable to some species of animals is demonstrated by the "cost" of anatomically growing, physiologically maintaining, and behaviorally displaying structures that have communication as their principal or, sometimes, their only known function. An extreme case is the one greatly enlarged claw of male fiddler crabs, which constitutes a third or more of the body weight in certain species. This structure is very rarely, if ever, used for anything but social communication—chiefly ritualized aggression between males, and courtship (Crane, 1975). Fiddler crabs and many other animals must be more vulnerable to predation than would otherwise be the case, because of conspicuous structures or conspicuous behavior involved in social communication.

Even more commonly, large amounts of time and energy are consumed in intermale aggression or courtship behavior than would seem at all necessary for the simple requirement of bringing together males and females ready to mate. These costs seem large compared to those proposed by evolutionary biologists to account for the evolution of morphological characters. For example, the trend for many species of birds and mammals to be larger in size and to have shorter extremities at higher latitudes usually is explained by postulating that the relatively slight decrease in surface-to-volume ratio reduces heat loss and thus conserves metabolic energy. While such things are difficult to measure, it seems likely that the added cost of competitive group displays at "leks" of grouse and other birds far exceed these differences in heat loss. Hence, according to the basic axioms of evolutionary biology, such social displays must have been favored by some selective advantage great enough to outweigh their cost. Wilson (1975) discusses these questions with many specific examples.

Communication behavior is probably most likely to resemble human language in species whose social behavior involves a high degree of interdependence, so that it is adaptively advantageous to have an efficient means of communication between individuals to coordinate their activities. But, since social communication is by no means limited to men and honeybees, versatile signaling systems should be advantageous to many species.

In a wide-ranging and stimulating discussion of the evolution of mental processes, Julian Huxley and Niko Tinbergen expressed a fundamental disagreement concerning the likelihood that animals have subjective mental experiences (Tax and Callender, 1960, pp. 175–206 and 267). Huxley held that they probably do, and that the question is a valid one, open to scientific investigation. Tinbergen argued the contrary position that we have no basis for inferring subjective experiences in other species. During this discussion, Huxley was asked whether conscious awareness is adaptive in the sense that this term is used by evolutionary biologists, that is, whether it has a survival value and hence has been favored by natural selection. Huxley was sure that it does, but his reasons were not stated in any detail in this symposium. Because of its importance, I should like to take up this question where Huxley left off and present arguments that awareness is indeed adaptive.

In strictly operational terms, awareness can be considered as readiness to respond to certain patterns of stimulation. But such indirect evidence may be unduly limited, or even misleading. For instance, an animal with only one operative sensory channel—an electric fish in muddy water, for example—could be subjectively aware of a simple, but important, communication signal, such as threat of attack conveyed by a change in the frequency of the electric discharges from another electric fish (Hopkins, 1974). Yet, in the absence of other information, such a fish would be operationally indistinguishable from an electronic frequency meter. Must we therefore define awareness in terms of the numbers and complexity of the signal patterns to which an animal is ready to respond appropriately, or is the criterion necessarily a subjective one?

An alternate approach is to consider awareness as the existence of internal images available for comparison with current sensory input. This recalls the cybernetic concept of a "Sollwert," the value of a sensory input which the animal tends to keep constant by adjustments of its behavior (Mittelstaedt, 1972). But this concept would have to be extended to include more than one sensory channel. Also related is the notion of a neural template (Marler, 1969). A sufficiently versatile template-matching machine, again in principle, could fulfill the *behavioral* criteria involved here. The psychological concept of a Gestalt is also applicable, at least in part. It is usually defined as a moderately complex pattern recognized from any of several viewpoints or when any of several redundant, overlapping stimulus patterns are perceived. The "middleness" concept, which chimpanzees and, to a much more limited extent, other primates have learned to recognize, is an example of such a perceived pattern (Rohles and Devine, 1966,

1967). Yet another example is provided by searching images, postulated internal images of something for which the animal searches (Croze, 1970).

The possession of mental images could well confer an important adaptive advantage on an animal by providing a reference pattern against which stimulus patterns can be compared; and it may well be an efficient form of pattern recognition. It is characteristic of much animal, as well as human, behavior that patterns are recognized not as templates so rigid that slight deviations cause the pattern to be rejected, but as multidimensional entities that can be matched by new and slightly different stimulus patterns, as when a familiar object is recognized from a novel angle of view. This ability to abstract the essential qualities of an important object and recognize it, despite various kinds of distortion, is obviously adaptive. Even greater adaptive advantage results when such a mental image also includes time as one of its dimensions, that is, the relationships to past and future events. Mental images with a time dimension would be far more useful than static searching images, because they would allow the animal to adapt its behavior appropriately to the probable flow of events, rather than being limited to separate reactions as successive perceptual pictures of the animal's surroundings present themselves one at a time. Anticipation of future enjoyment of food and mating or fear of injury could certainly be adaptive, by leading to behavior that increases the likelihood of positive reinforcement and decreases the probability of pain or injury.

The concept of mental images that include both spatial and temporal dimensions, like Skinner's definition of thinking as covert verbal behavior, tends to approach a working definition of conscious awareness. For instance, the image of food within reach might well be coupled with an image of the act of grasping the food, another of swallowing it, or even the image of its pleasant taste. Thus, if the existence of mental images in animals can be accepted as plausible, one need only postulate an appropriate linkage between them to sketch out a working definition of conscious awareness. It may be helpful, and even parsimonious, to assume some limited degree of conscious awareness in animals, rather than postulating cumbersome chains of interacting reflexes and internal states of motivation. Behavior patterns that are adaptive in the evolutionary biologist's sense may be reinforcing in the psychologist's terms, as well. Perhaps natural selection has also favored the mental experiences accompanying adaptive behavior.

It thus becomes almost a truism, once one reflects upon the question, that conscious awareness could have great adaptive value, in the sense that this term is used by evolutionary biologists. The better an animal understands its physical, biological, and social environment the better it can adjust its behavior to accomplish whatever goals may be important in its life, including those that contribute to its evolutionary fitness. The basic assumption of contemporary behavioral ecology and sociobiology, in the sense that the latter term is used by Wilson (1975) and many others, is that behavior is acted upon by natural selection along with morphological and physiological attributes. From this plausible assumption it follows that—insofar as any mental experiences animals have are significantly interrelated with their behavior—they, too, must feel the impact of natural selection. To the extent that they convey an adaptive advantage on animals, they will be reinforced by natural selection.

Arguments of this kind, which appeal to a presumed selective advantage, suffer from the limitation that a sufficiently fertile imagination can almost always find a plausible adaptive advantage for any trait which is observed. The very success of such arguments tends to undermine their strength, because if one can make up an equally plausible case for alternate explanations, there is little basis for preferring one explanation to any of the others. A stronger form of evolutionary argument is the converse position that any attribute with a selective *dis*advantage will almost certainly be eliminated unless it is genetically coupled with some compensating advantage. On this basis, we can at least argue that no selective drawbacks to conscious awareness have been demonstrated. Indeed, I am not aware that any have even been suggested.

CHAPTER

7

A Possible Window
on the Minds of Animals

THE AIM OF THIS FINAL CHAPTER is to outline potential experiments that may offer some realistic hope of escaping from the difficulties discussed in Chapters 4 and 5, and of beginning the exploration of scientific territory so unknown that even its existence has been seriously questioned. Since even thinking about mental experiences in animals has been largely tabooed throughout most of the development of ethology, it is not surprising that we have neither theoretical background nor empirical data with which to take up what I believe to be an important question. Furthermore, there are obvious dangers in postulating specific mental experiences in animals on the basis of their observed behavior when our only point of departure is to try to imagine our own mental experiences if we were to find ourselves in the animal's situation. The apparent hopelessness of proceeding effectively into this difficult area has been one of the major reasons for neglecting the questions raised in previous chapters.

A helpful point of departure is to compare the approach of an ethologist studying the communication behavior of another species with that of a hypothetical anthropologist making initial contact with a group of people whose language is totally unknown to him (Nance, 1975°). Because they are men, the anthropologist assumes that their sounds are indeed a form of speech. He notes correlations between their behavior patterns and their vocalizations and gestures, and he is certain to rely heavily on gestures in his own first attempts to communicate. The importance of nonverbal human communication is receiving increased recognition and is being effectively investigated from many viewpoints (Hinde, 1972; Krames et al., 1974). Our hypothetical anthropologist might take advantage of his knowledge of this field to encourage the people with whom he is trying to communicate to make at least some effort in the same direction, perhaps by pointing to persons or objects while uttering their names. Curiously enough, linguists and anthropologists have paid very little attention to the first steps needed to establish linguistic contact with people speaking languages unknown to one another (Hewes, 1974, 1975°).

It is conceivable that a diligent anthropologist might learn a great deal about a language by one-way visual and auditory observation of people as they use it. In principle, the same thing might be done by watching extensive television or motion-picture sequences provided with an adequate sound track. But even if success should be claimed from such a laborious effort, we would wish to test the claim by asking that direct, two-way communication be demonstrated. Indeed, we would have rather little confidence that the anthropologist had really learned the language until his knowledge passed this crucial test.

Studies of animal communication have so far remained almost entirely at a comparable level of correlated observations, except for the recent studies of chimpanzees. We see that animals emit certain signals and observe how conspecifics respond. The signals may be sounds, visual patterns, scents, tactile vibrations, or even electric currents (Hopkins, 1974), and each channel presents different problems of monitoring what are suspected to be signals, or playing back artificial signals to observe any responses they may elicit. But except for teaching sign language to apes, scarcely any effort has been made to move ahead to the next stage of *participatory* investigation. The Gardners succeeded in establishing far more complex two-way communication with Washoe than had any of their predecessors. One element in their success was the use of a communication channel (manual gestures) that turned out to be more appropriate for chimpanzees than vocal sounds. They and Washoe also had the advantage of being enough alike morphologically so that acceptance as partners in social-communication behavior was easier than between, say, man and dog. Yet millions of pet lovers have achieved limited forms of communication with dogs, cats, and other animals. The complexity and versatility of such communication is obviously very limited, compared to the common vocabulary established between the Gardners and Washoe or between other investigators and other chimpanzees.

PARTICIPATORY INVESTIGATION OF
ANIMAL COMMUNICATION

I should like to suggest that it is now appropriate and promising to extend these approaches to other species, using methods analogous to those of anthropologists seeking to establish communication with conspecifics who are assumed to speak some language, but one in which no words are yet shared. It will be necessary, almost literally speaking, to talk back and forth with a communicating animal in order to verify the meanings of its communication signals. Most animals are sufficiently different from men that an investigator is unlikely to be an acceptable social partner. In these cases, appropriate models are called for. Such models must be similar enough to the animals in question, and sufficiently versatile in their signaling behavior, to act as transmitters of information via whatever communication system is natural to the animals under study. Initially, the investigator would act as the receiver, through appropriate observations, but at a later stage in the development of such experiments he could manipulate the model to attempt *two-way communication*. Symbolic communication might, if suitably developed, provide us with a "window" through which to examine the properties of an animal's neural templates, Gestalten, or mental images. The data gathered in this way should, of course, be validated by comparison with conventional stimulus-response experiments.

Many animals respond well enough to various types of models to offer realistic hopes of eventual success (reviewed by Tinbergen, 1951; and by Marler and Hamilton, 1966). For example, some invertebrate and vertebrate animals react to their mirror images. In most cases, this does not lead to any prolonged interaction, but the mere fact that mirror images elicit any communication behavior at all tells us that visual signals can be mimicked with at least limited success.

Stout and Brass (1969) and Stout, Wilcox, and Creitz (1969) have demonstrated that some important elements of aggressive display between glaucous-winged gulls (*Larus glaucescens*) can be elicited by relatively simple models with or without playback of vocalizations used in normal aggressive encounters. But these experiments were not carried past the initial stage of showing that certain sounds or postures elicited stronger responses than others. The experiments of Chauvin-Muckensturm (1974), in which woodpeckers learned to ask for preferred foods by a simple telegraphic code, suggest that birds may have greater capabilities for symbolic communication than have yet been recognized.

Sounds are much easier to monitor, record, and reproduce than are visual or chemical signals, and playback experiments have amply demonstrated their value in analyzing a wide variety of acoustical communication systems, such as the territorial calls exchanged between neighboring males in many species of birds. Such birds can easily be induced to exchange territorial calls with a tape recorder, but relatively little effort has yet been devoted to attempts at a more detailed two-way communication, comparable to the gestural communication established between the Gardners and Washoe. An important challenge is to analyze the nuances

and details of animal communication, to inquire whether messages more subtle than the gross assertions of territoriality, for
example, are exchanged. It is even easier than with birds to
establish simple "dialogues" with electric fishes, because crude
models provided with appropriate electrodes can readily emit and
monitor the electrical communication signals (Hopkins, 1974;
Westby, 1974). The possibility of communicating with a conspecific or with a model or mirror image sometimes appears to be
reinforcing, that is, it seems that some animals like to engage in
this kind of activity (Stevenson, 1969). Birds can certainly
recognize each other as individuals through details of song pattern
(W. J. Smith, 1969; Falls, 1969; Falls and Krebs, 1975°), so that
the basic requirements for elaboration of more detailed communication are clearly present. W. J. Smith (1969) and Wilson (1975)
find only a relatively small number of distinguishable messages,
but perhaps a closer scrutiny via attempts at two-way communication would disclose a finer grain in the process, or significant
combinations of messages.

Beer (1975, 1976) has recently discovered just this kind of
"finer grain" in the communication behavior of laughing gulls
(*Larus atricilla*), including both sounds and visual signals conveyed by postures and motions. The long-call, which had
previously been interpreted as a single communication signal,
turns out to convey different messages under various circumstances, a case of context-dependence of the kind discussed by W.
J. Smith (1968, 1969°). The long-call also serves in some cases to
identify the individual gull, so that it is recognized by its mate,
neighbors, and chicks. Quite independent of the development of
the ideas expressed in this book, Beer was led by his discoveries of
the remarkable complexity and versatility of gull communication
to conclude that "the long-call of Laughing Gulls is a form of
display by means of which a gull can emphatically identify itself
and convey a number of alternate messages with regard to itself . . . a long-call might thus signify 'I am your parent—come
and get fed'; or 'I am your mate—let me sit on the eggs'; or 'I am

your prospective mate—come and stay close'; or 'I am the occupier of this area—get out'. ... the long-call ... is semantically and pragmatically open'" (Beer, 1975). Elsewhere, Beer (1976) writes: "the recognition of greater complexity has resulted in, and in turn caused, changes in preconceived views about animal communication, including the models in terms of which animal communication has been thought about ... linguistic analogies have, to some extent, taken the places previously occupied by causal and statistical models."

The dance-communication system of honeybees provides another significant example, in which both the promise and the difficulties can readily be appreciated. Since men and bees are so enormously different in size and morphology, to say nothing of gesturing capabilities, we need to develop an effective model bee. An important step in this direction was taken some years ago by Esch (1964), and J. L. Gould (personal communication) has made substantial improvements on that simple model. But much remains to be done before we will have available a model that can both emit appropriate gestures and be accepted by the bees so that they act in accordance with the information transmitted from the model. A further step would be to use the model as a recipient, that is, to elicit dancing behavior in a real bee. Enough is already known so that an experimenter could observe and interpret the dances directed at a model whose behavior as a recipient of information could be controlled experimentally.

A successful model bee must not only display the correct mechanical motions, but also the appropriate odors. The latter may not yet be well-enough known to permit adequate simulation, but this gap could be closed by sufficient investigation. Sounds and mechanical vibrations are also important, and these, too, may have to be simulated with more precision than any experimenter has yet achieved. Gould has already perfected another important aspect of such a model—the provision of an artificial substitute for regurgitation of sugar solution from the honey stomach. His model has been developed to the point at which other bees have accepted sugar solution from its artificial proboscis, although it is not yet clear whether the chemical stimulation resulting from this artificial trophallaxis is at all normal. The technical requirements for a truly adequate model honeybee are thus formidable, but by no means impossible. Such efforts, not only with honeybees, but with other animals, as well, are extremely worthwhile, and highly significant results could be obtained from intensive and thoughtful effort.

Before 1945, it was easy to observe honeybees in an observation hive under circumstances where a great deal of dancing was taking place and yet fail to recognize its significance. Numerous bees were moving in rapid and complex patterns, but this showed only that they were aroused. Having the benefit of von Frisch's insights, we can now see a complex but orderly communication process at work. But suppose we continue the investigation in more detail and study more closely the dancing of a single bee. Not all the dances are precisely alike; the direction varies within limits, and so does the duration of the straight run that signals distance. There is also a quality of vigor, or enthusiasm, which is evident to the casual observer and which has not been defined in a wholly adequate fashion. One of its properties is "duty cycle"— the fraction of her time which a returning forager spends in dancing. If she is bringing back a very rich sugar solution when the colony is in need of carbohydrates, she dances a large fraction of the time, and each bout of dancing is repeated for many cycles. If the nectar or artificial sugar solution is very dilute, she may dance only a small fraction of the time and each bout may include only one or two cycles. But other properties of the dances related to the amplitude or the specific patterns of motion by the abdomen of the dancer probably convey her enthusiasm, because it seems that a bee can receive the message by observing and following only a few dances. The responses of other bees seem also to be influenced by the acoustical complex of airborne sounds and substrate vibrations; for example, dances not accompanied by sounds are ineffective in recruiting other bees to fly out to the location signaled by the dancer.

So far, these have been conservative descriptive statements, comparable to the pre-1940 observation that, under certain circumstances, many bees return to a hive and carry out complex and variable motions of no known significance. Let us entertain for the moment the hypothesis that the dancing conveys additional information, perhaps about the nature of the food source, along dimensions other than a single linear scale of desirability. If there is a more extensive and finer-grained two-way communication going on, how likely would we be to discover it by the type of observation customary to date? If, on the other hand, a model bee could be successful enough to be accepted by the real bees as a partner in communication behavior, we might learn about the existence and nature of more subtle elements in the communication system by participatory experiments analogous to the extensive and complex social interactions between the Gardners and Washoe.

This whole approach to the study of communication behavior in animals is so poorly developed, even at the level of preliminary hypotheses, that it is difficult to anticipate where it might lead. Perhaps the outcome of numerous and laborious experiments would be a negative result, in that no additional kinds or degrees of communication would be discovered. For example, von Frisch (1967) examined the possibility that bees might convey information about vertical, as well as horizontal, directions, but obtained a convincing negative answer. Even such results would be of value in setting more precisely defined limits to the communicative capabilities of the species in question. Alternately, however, it might be discovered that previously unsuspected messages are exchanged, and the scientific interest of such discoveries is self-evident. If we consider the recent history of this field, it is clear that far more complex communication behavior has been found than any scientist would have ventured to predict 30 years ago. Have we any reason to believe that progress in this exciting area has reached a sudden end? We do not know whether the properties of various animal communication systems, as they are now understood, are limited by the capabilities of the communication systems themselves or by the methods of investigation that have so far been employed.

One challenging approach would be direct "impersonation" of a similar species, such as a chimpanzee, by an adequately disguised experimenter using the gestures and sounds characteristic of chimpanzee communication. The disguise would have to be thorough, including not only visual appearance, but also chimpanzee sounds and appropriate pheromonal perfumes. Jane Goodall (1971) approached a state of acceptance by wild chimpanzees without any attempt at morphological disguise. A new generation of ambitiously pioneering ethologists might open up an enormously powerful new science of participatory research in interspecies communication. First, they must overcome the feeling of embarrassed outrage at this notion, and then laboriously develop the necessary techniques of disguise, imitation, and communicatory interaction.

TOWARD A COMPARATIVE LINGUISTICS

Animal surrogates have been invaluable in the analysis and explanation of many biological phenomena, including some aspects of behavior, such as learning. The resulting knowledge and understanding has had many important applications to human medicine in particular and human affairs in general. Biologists have often found that particular phenomena are more easily or more effectively investigated in certain species, a principle often attributed to the physiologist August Krogh (Krebs, 1975). When a biological process is known only in our own species, its investigation is often difficult or impracticable. But when one or more appropriate animal surrogates are discovered, many kinds of experiments become feasible and scientific understanding is more readily attainable. It is common knowledge that cures for many human diseases have been discovered by this general procedure, and human welfare has been served in many other areas, such as in the study of nutrition. It is also obvious that this basic approach depends heavily on evolutionary continuity and the resulting confidence that the same basic principles can be applied to animal and human physiology.

Social communication behavior, broadly defined, is clearly of the highest importance in human affairs, comparable in importance to nutrition and physical health. But, unfortunately, we do not understand it nearly so well as many other areas of biology. This, in turn, suggests that experimental analyses of social-communication behavior making use of animal surrogates could contribute significantly to a better understanding of human psychology, sociology, and even such apparently nonbiological disciplines as economics and philosophy. Of course, such comparative analyses can never do the whole job, and any resulting conclusions must be checked against data obtained by studying human beings—just as new drugs or biochemical processes developed by animal experiments need careful checking with human subjects before their general application is wise. But many advances in the biomedical sciences would have been greatly impeded, if not hopelessly crippled, without the use of animal surrogates for basic research.

In communication behavior, including human language, our current situation is needlessly hampered by a tendency to deny *a priori*, on theoretical grounds, that evolutionary continuity exists and that animal surrogates are thinkable. The appropriate analogy is to a nutritional problem which people were so convinced was a purely human phenomenon that they refused to test potential dietary supplements on experimental animals. If one believes with Chomsky (1966) that human language is "based on an entirely different principle," a whole avenue of investigation is blocked off. This is surely an inefficient approach to an important and challenging cluster of scientific problems. On the other hand, to the extent that evolutionary continuity is significant in communication behavior, the entire momentum of comparative, experimental science can be brought to bear on what has previously appeared to be a uniquely specialized and almost unapproachable phenomenon.

Participatory two-way experiments of the general sort discussed above hold the potential of revealing important properties of animal-communication systems that could be explored only slowly, and with many uncertainties, by correlating signals the animals exchange with simultaneously or subsequently observed behavior. The immediate advantages lie in the possibility of controlling the messages experimentally in order to determine their effective content. This line of inquiry should, in due course, enable ethologists to work out just what messages are, in fact, exchanged under various conditions between animals of a given species. But, in addition to this near objective, there is also a more important, though distant, hope that such methods will permit first the detection and later the analysis of any cognitive processes that occur in the brains or minds of animals. This means that a most important step toward understanding communication behavior will come when an investigator can ask questions and receive answers about any possible mental experiences (or, if behaviorists prefer, about covert communication behavior) in a given animal. Human language, despite all its limitations, does convey some information about subjective experiences in our fellow men. Improved methods of investigating communication behavior in animals might open up comparable avenues of investigation into their subjective experiences, if such exist. In the case of Washoe and other chimpanzees, simple questions about desires for the near future have been asked by the human investigator and answered by the chimpanzee. Bees seem to ask and answer questions about the desirability and location of food or new hive sites. We need only extend to animals, with appropriate modifications, the basic process by which we assess the mental experiences of our fellow men.

One pessimistic criticism of the aspirations sketched above will no doubt take the form suggested by Wittgenstein (1953): "If a lion could talk we could not understand him." Human thoughts and words are closely linked, at the very least, and, as has been discussed in Chapter 3, many philosophers have argued that they are essentially identical. This argument assumes that human mental experiences are so closely bound up with our species-specific neurophysiological mechanisms that we are not capable of understanding any mental, as distinct from neurophysiological, processes in other animals, even if such exist. In this view, should other species have feelings, hopes, plans, or concepts of any sort—even very simple ones—they would take a form so different from our own thoughts that we could not recognize them. But physiologists have found that more and more of the basic properties and functions of neurons and nervous systems are remarkably similar in virtually all multicellular animals. This is consistent with the assumption of evolutionary continuity, leading us to use animal surrogates in studying the functioning of the nervous system, even with reference to problems of human brain function.

As discussed by Premack (1975) and by Marler (in prep.), much animal communication conveys an emotional or affective state, such as fear. Except for the dances of honeybees, and the sign language recently taught to captive chimpanzees, it is difficult to judge from available data whether animal communication behavior also includes specific information about the nature of the object or situation responsible for the emotional state conveyed. But rather than assuming the absence of such information *a priori*, it seems advisable to consider this an open question to be investigated. For several decades, behavioral scientists have strongly tended to neglect such questions as either meaningless or hopelessly difficult and, in the process, we have tended to drift into the unwarranted assumption that subjective experiences are nonexistent. This constitutes a great change from the viewpoint of Charles Darwin (1872), who took it for granted that animals had mental experiences and emotions, and devoted one of his major books to the behavior patterns by which animals expressed those emotions.

A major objection to the return of this Darwinian viewpoint is that postulating mental experiences in animals does not immediately suggest definitive experiments that can decide between alternate hypotheses. Another version of this criticism is that no falsifiable hypotheses have been presented, that is, hypotheses which are susceptible to experimental tests leading to a firm verdict of true or false. By this is meant that alternate and almost, if not quite, equally plausible interpretations are reasonably drawn from whatever outcome the proposed observations or experiments might yield. Thus, I may interpret the phenomena of bee dances as evidence that workers intentionally and consciously communicate information, whereas a classical behaviorist may argue with equal force that the bees are complex automata, that the dances are correlates of certain physiological states, but that no conscious intention need be assumed to exist merely because symbolic communication is taking place. How can we work our way out of this dilemma?

One avenue of partial escape is to expand a suggestion advanced in Chapter 4 by pointing out that many areas of science have made substantial and significant progress by formulating relatively imprecise hypotheses, gathering evidence pertinent to them, and gradually building up a persuasive, if not meticulously rigorous, case for an important conclusion. Charles Darwin operated in essentially this manner, along with a host of his followers. The resulting conviction that men are genetically descended from animals had as great an impact on human thinking as any other scientific discovery. Only the Copernican revolution is a close rival, but altering fundamental views about the heavens was not as basic a change in our outlook on the universe as the recognition that we ourselves are genetically related to animals. Established religions vigorously resisted both revolutions, but later accepted the newly recognized facts and adjusted their beliefs accordingly. In the case of biological evolution, that process is still not complete, and the issues discussed in this book are significantly debatable as part of this accommodation.

In other scientific disciplines, new concepts or entities are proposed to help explain observations or the results of experiments without such adamant requirements for precise verification as those demanded by strict behaviorists for hypotheses about mental experiences. The well-known dilemma of physicists concerning the wave and quantal properties of electromagnetic radiation have not led them to stop all investigations of quantum mechanics or particle physics, simply because they cannot yet tell us whether light is waves or particles or explain how it can have the properties of both at the same time. Likewise, paleontologists do their best to make sense out of the fossil record and sketch in evolutionary sequences or unfossilized morphologies without realistic hope of obtaining specific verification within the foreseeable future.

Why must we be so insistent that mental experiences must be delivered in neatly concrete form, like an album of postage stamps, with all important particulars totally visible and susceptible to repeated and consistent measurement of every detail, before we allow ourselves to consider the possibility that they have any significant reality? Behaviorists seem content only with what might be called true-false or, at best, "multiple choice" experiments.

Even if the above arguments are accepted, however, there is no doubt that any scientist would be more satisfied if the concepts with which he deals are as fully observable and verifiable as possible—preferably more so than either neutrinos or mental images are at present. This is why I have based this discussion on human mental experiences, which most of us (even the strict behaviorists, when forced into a corner) recognize as real and significant. Language or, in a broader sense, communication behavior, is virtually the only method we have for learning about the subjective mental experiences of our fellow men. Furthermore, as discussed in detail above, the newly recognized versatility of animal communication behavior opens up a strictly comparable window by which we can hope to learn something, at least, about whatever mental experiences they may have. In other words, the possibility of animal introspection is more than a will-o'-the-wisp; it is a potential method which has already been employed to a very limited degree by the Gardners, Fouts, and other students of chimpanzees, and one that is ready for development and exploitation with other species to roughly the degree that they employ flexible communication systems. Clearly, the more versatile the communication system, which probably, but not necessarily, means the more symbolic it may be, the greater the opportunities for ethologists to employ it as a source of information about whatever mental images, intentions, and awareness such animals may have.

The evidence and ideas reviewed above lead to the cautiously stated conclusion that complex, versatile, and adaptive responses may or may not be accompanied by conscious awareness. How can we hope to find out? If an animal communicates about its internal images, this provides at the very least some data about them, but it does not, strictly speaking, tell us conclusively whether the animal is consciously aware of those images or other relationships. But the general principle of evolutionary kinship and continuity between animals and men suggests that an animal which communicates about its internal images may also be aware of them on some occasions. The hypothesis that some animals are indeed aware of what they do, and of internal images that affect their behavior, simplifies our view of the universe by removing the need to maintain an unparsimonious assumption that our species is qualitatively unique in this important attribute. By concentrating on communicating animals in the hope of learning something about their mental experiences, we should not overlook the broader possibility that some animals may also have mental experiences about which they do not communicate. Thus, we may be able for the forseeable future to discern only the tip of a large and significant iceberg. But this is preferable to intentionally holding before our eyes an "iceberg-rejecting filter."

The pessimistic view that we are inherently incapable of communicating with animals involves the same sort of patronizing discouragement that I recall so vividly with regard to sun- and star-compass orientation or the use of echolocation to capture flying insects. A more hopeful prospect is based on the success of comparative physiologists, comparative psychologists, and ethologists studying animal orientation and communication in analyzing processes and mechanisms whose very existence was undreamed-of until very recently. Of course, to recognize the promise of a new approach is only the barest beginning. But the road now seems open, and I doubt that it will turn out to be a dead-end street. It will be of the greatest interest and significance if it should become increasingly likely that animals have mental experiences, however few in number and simple in nature they turn out to be. This important step might then lead us into effective investigation of what may properly be called cognitive ethology.

This chapter has emphasized the potential value of animal communication in general, and two-way participatory experiments in particular, as a source of information about whatever mental experiences animals may have. Of course, this is not the *only* source of evidence about the internal workings of animal brains or minds. Many other categories of versatile behavior can also be interpreted plausibly in terms of conscious intentions on the animals' part, for instance, constructing and using simple tools, building shelters or otherwise improving the immediate environment, aiding injured companions, sharing food, or hunting cooperatively. Many scientists who work with primates or dogs recognize the likelihood of at least some conscious awareness. Indeed, one response to the ideas outlined in this book has been "Why, yes, of course." As behavioral scientists recognize more clearly and widely the probable existence of mental experiences, we can proceed far more effectively to study their nature, causation, and significance.

CHAPTER

8

Summary and Conclusions

THE COMMUNICATION BEHAVIOR of certain animals is complex, versatile, and, to a limited degree, symbolic. The best-analyzed examples are the dances of honeybees and the use of American Sign Language by several captive chimpanzees. These and other animal communication systems share many of the basic properties of human language, although in very much simpler form.

Language has generally been regarded as a unique attribute of human beings, different in kind from animal communication. But on close examination of this view, as it has been expressed by linguists, psychologists, and philosophers, it becomes evident that one of the major criteria on which this distinction has been based is the assumption that animals lack any conscious intent to communicate, whereas men know what they are doing. The available evidence concerning communication behavior in animals suggests that there may be no qualitative dichotomy, but rather a large quantitative difference in complexity of signals and range of intentions that separates animal communication from human language.

Human thinking has generally been held to be closely linked to language, and some philosophers have argued that the two are inseparable or even identical. To the extent that this assertion is accepted, and insofar as animal communication shares basic properties of human language, the employment of versatile communication systems by animals becomes evidence that they have mental experiences and communicate with conscious intent. The contrary view is supported only by negative evidence, which justifies, at the most, an agnostic position.

According to the strict behaviorists, it is more parsimonious to explain animal behavior without postulating that animals have any mental experiences. But mental experiences are also held by behaviorists to be identical with neurophysiological processes. Neurophysiologists have so far discovered no fundamental differences between the structure or function of neurons and synapses in men and other animals. Hence, unless one denies the reality of human mental experiences, it is actually parsimonious to assume that mental experiences are as similar from species to species as are the neurophysiological processes with which they are held to be identical. This, in turn, implies qualitative evolutionary continuity (though not identity) of mental experiences among multicellular animals.

The possibility that animals have mental experiences is often dismissed as anthropomorphic because it is held to imply that other species have the same mental experiences a man might have under comparable circumstances. But this widespread view itself contains the questionable assumption that human mental experiences are the only kind that can conceivably exist. This belief that mental experiences are a unique attribute of a single species is not only unparsimonious; it is conceited. It seems more likely than not that mental experiences, like many other characters, are widespread, at least among multicellular animals, but differ greatly in nature and complexity.

Awareness probably confers a significant adaptive advantage by enabling animals to react appropriately to physical, biological, and social events and signals from the surrounding world with which their behavior interacts.

Opening our eyes to the theoretical possibility that animals have significant mental experiences is only a first step toward the more difficult procedure of investigating their actual nature and importance to the animals concerned. Great caution is necessary until adequate methods have been developed to gather independently verifiable data about the properties and significance of any mental experiences animals may prove to have.

It has long been argued that human mental experiences can only be detected and analyzed through the use of language and introspective reports, and that this avenue is totally lacking in other species. Recent discoveries about the versatility of some animal communication systems suggest that this radical dichotomy may also be unsound. It seems possible, at least in principle, to detect and examine any mental experiences or conscious intentions that animals may have through the experimental use of the animal's capabilities for communication. Such communication channels might be learned, as in recent studies of captive chimpanzees, or it might be possible, through the use of models or by other methods, to take advantage of communication behavior which animals already use.

The future extension and refinement of two-way communication between ethologists and the animals they study offer the prospect of developing in due course a truly experimental science of cognitive ethology.

Bibliography

BIBLIOGRAPHY

ADAMS, D. K. 1928. The inference of mind. *Psychol. Rev.* 35: 235–252.

ADLER, M. J. 1967. *The Difference of Man and the Difference it Makes.* New York: Holt, Rinehart and Winston.

ALSTON, W. P. 1972. Can psychology do without private data? *Behaviorism* 1: 71–102. The conclusion is that it can.

ANSHEN, R.N. 1957. Language as Idea. In: Anshen, R. N. (Ed.), *Language: An Enquiry into its Meaning and Function.* New York: Harper.

APTER, M. J. 1970. *The Computer Simulation of Behavior.* London: Hutchinson University Library. A thoughtful discussion of the relationship of computer capability to consciousness.

ARMSTRONG, E. A. 1949. Diversionary display. *Ibis* 91: 88–97 and 179–188.

AYALA, J., AND DOBZHANSKY, T. (Eds.). 1974. *Studies in the Philosophy of Biology, Reduction and Related Problems.* Berkeley: Univ. of California Press.

BEER, C. G. 1973a. Behavioral Components in the Reproductive Biology of Birds. In: *Breeding Biology of Birds.* Washington, D.C.: National Academy of Sciences, pp. 323–365.

———. 1973b. A View of Birds. In: Pick, A. (Ed.), *Minnesota Symposium on Child Biology.* Vol. 7. Minneapolis: Univ. of Minnesota Press.

———. 1975. Multiple Functions and Gull Displays. In: Baerends, G., Beer, C. G., and Manning, A. (Eds.), *Essays on Function and Evolution in Behaviour: A Festschrift for Professor Niko Tinbergen.* Oxford: The Clarendon Press, Chapter 2.

———. 1976. Some Complexities in the Communication Behavior in Gulls. *Proc. of Conference on the Origins and Evolution of Language. Ann. N.Y. Acad. Sci.* (in press).

BENNETT, J. 1964. *Rationality, Essay towards an Analysis.* London: Routledge and Kegan Paul.

BERG, P., et al. 1974. Potential biohazards of recombinant DNA molecules. *Science* 185: 303.

BERTRAND, M. 1969. *The Behavioral Repertoire of the Stumptail Macaque, A Descriptive and Comparative Study.* Basel: S. Karger (Bibliotheca Primatologica No. 11).

BLACK, M. 1968. *The Labyrinth of Language.* New York: Praeger.

BLOOMFIELD, L. 1933. *Language*. New York: Holt (Reprinted 1961 by Holt, Rinehart and Winston). Bloomfield recognized that animals do communicate but concluded: "Human speech differs from the signal-like actions of animals, even of those which use the voice, by its great differentiation. Dogs, for instance, make only two or three kinds of noise—say, barking, growling, and whining. . . . When we tell someone, for instance, the address of a house we have never seen, we are doing something which no animal can do."

BORING, E. G. 1963. *The Physical Dimensions of Consciousness*. New York: Dover Publications.

BOYLE, D. G. 1971. *Language and Thinking in Human Development*. London: Hutchinson, pp. 29, 37, 152–153.

BREWER, W. F. 1974. There is No Convincing Evidence for Operant or Classical Conditioning in Adult Humans. In: Weimer, W. B., and Palermo, D. S. (Eds.), *Cognition and the Symbolic Processes*. Hillsdale, N. J.: Lawrence Erlbaum Associates.

BRONOWSKI, J., AND BELLUGI, U. 1970. Language, name, and concept. *Science* 168: 669–673. The authors accept the early data from Washoe as demonstrating naming, but conclude that she does not use sentences and that her signing differs from human language in lacking rule-guided sequences of words.

BROWN, J. L. 1975. *The Evolution of Behavior*. New York: Norton.

BROWN, R. 1958. *Words and Things*. New York: Free Press of Glencoe.

BROWN, R. G. B. 1962. The aggressive and distraction behaviour of the western sandpiper *Ereunetes mauri*. *Ibis* 104: 1–12.

BROWN, S. R. 1972. A Fundamental Incommensurability between Objectivity and Subjectivity. In: Brown, S. R., and Brenner, D. J. (Eds.), *Sciences, Psychology, and Communication. Essays honoring William Stephenson*. New York: Teachers College Press.

BULLOCK, T. H. 1973. Seeing the world through a new sense; electroreception in fish. *Am. Sci.* 61: 316–325.

CASSIRER, E. 1953. *The Philosophy of Symbolic Forms*. Vol. 1. *Language*. New Haven: Yale Univ. Press, p. 189 ff. "Even among the lower animals we encounter a great number of original sounds expressing feeling and sensation . . . cries of fear or warning, lures or mating calls. But between these cries and the sounds of designation and signification characteristic of human speech there remains a gap, a 'hiatus' which has been newly confirmed by sharper methods of observation of modern animal psychology." Here Cassirer cites Köhler (1925) in support of his conclusion that "The step to human speech, as Aristotle stressed, has been taken only when the purely significatory sound has gained primacy over sounds of affectivity and stimulation. . ."

CHAUVIN-MUCKENSTURM, B. 1974. Y a-t-il utilisation de signaux apprix comme moyen de communication chez le pic epeiche? *Rev. Comp. Animal* 9: 185–207.

CHOMSKY, N. 1959. Review of Skinner's *Verbal Behavior. Language* 35: 26–58. Reprinted with comments in Jacobovits, L. A., and Miron, M. S. (Eds.), *Readings in the Psychology of Language.* Englewood Cliffs, N.J.: Prentice-Hall.

———. 1966. *Cartesian Linguistics.* New York: Harper & Row.

———. 1972. *Language and Mind.* New York: Harcourt Brace Jovanovich.

COWGILL, V. M. 1972. Death in *Perodicticus. Primates* 13: 251–256.

CRANE, J. 1975. *Fiddler Crabs of the World.* Princeton: Princeton Univ. Press.

CRITCHLEY, M. 1960. The Evolution of Man's Capacity for Language. In: Tax, S. (Ed.), *Evolution after Darwin.* Vol. II. Chicago: Univ. of Chicago Press.

CROZE, H. 1970. Searching image in carrion crows. *Tierpsychol.* Suppl. 5. Berlin: Parey.

DAANJE, A. 1951. On the locomotory movements in birds and the intention movements derived from them. *Behaviour* 3: 48–98.

DARWIN, C. 1872. *The Expression of the Emotions in Man and Animals.* London: Murray. (Reprinted 1965 University of Chicago Press, with Preface by Konrad Lorenz).

DOBZHANSKY, T. H. 1967. *The Biology of Ultimate Concern.* New York: New American Library.

ECCLES, J. C. 1973. Brain, speech and consciousness. *Naturwissenschaften* 60: 167–176.

———. 1974. Cerebral Activity and Consciousness. In: Ayala, F. J., and Dobzhansky, T. (Eds.), *Studies in the Philosophy of Biology, Reduction and Related Problems.* Berkeley: Univ. of California Press.

ELITHORN, A., AND JONES, D. (Eds.). 1973. *Artificial and Human Thinking.* San Francisco: Jossey-Bass, and Amsterdam: Elsevier.

EMLEN, S. T. 1967. Migratory orientation in the Indigo Bunting, *Passerina cyanea. Auk* 84: 309–342 and 463–489.

———. 1975. Migration: Orientation and Navigation. In:Farner, D. S., and King, J. R. (Eds.), *Avian Biology.* Vol. V. New York: Academic Press.

ESCH, H. 1961. Über die Schallerzeugung beim Werbetanz der Honigbiene. *Z. vgl. Physiol.* 45: 1–11.

———. 1964. Beiträge zum Problem der Entfernungsweisung in den Schwänzeltänzen der Honigbiene. *Z. vgl. Physiol.* 48: 534–546.

ESCH, H., AND BASTIAN, J. A. 1970. How do newly recruited honey bees approach a food site? *Z. vgl. Physiol.* 68: 175–181.

ESTES, W. K. (Ed.). 1975. *Handbook of Learning and Cognitive Processes.* Vol. 1. *Introduction to Concepts and Issues.* Hillsdale, N.J.: Lawrence Erlbaum.

FALLS, J. B. 1969. Functions of Territorial Song in the White-throated Sparrow. In: Hinde, R. A. (Ed.), *Bird Vocalizations.* London: Cambridge Univ. Press.

FALLS, J. B., AND KREBS, J. R. 1975. Sequences of songs in repertoires of Western Meadowlarks (*Sturnella neglecta*). *Can. J. Zool.* 53: 1165–1178. Nonrandom responses to playback, conservatively analyzed.

FEIGL, H., SELLARS, W., AND LEHRER, K. 1972. *New Readings in Philosophical Analysis.* New York: Appleton-Century-Crofts. Here in a volume with scholarly, though divergent, views by several leading philosophers, Feigl maintains that the mind-body problem is not a pseudoproblem.

FEYERABEND, P. K., AND MAXWELL, G. 1966. Mind, Matter, and Method. *Essays in the Philosophy of Science in Honor of Herbert Feigl.* Minneapolis: Univ. of Minnesota Press.

FODOR, J. A. 1968. *Psychological Explanation, An Introduction to the Philosophy of Psychology.* New York: Random House. A forceful attack on the behavioristic view of psychology, including: "Finally, it seems evident that behaviorism, considered as an account of theories of psychology, is simply a special case of operationalism considered as an account of scientific theories in general and that the former doctrine ought therefore to share in the discredit that has recently attached to the latter . . . the ultimate argument against behaviorism is simply that it seeks to prohibit *a priori* the employment of psychological explanations that may, in fact, be true."

———. 1975. *The Language of Thought.* New York: Thomas Y. Crowell. Here Foder argues further for the reality of mental processes.

FODOR, J. A., BEVER, T. G., AND GARRETT, M. F. 1974. *The Psychology of Language.* New York: McGraw-Hill.

FOUTS, R. S. 1973. Capacities for Language in the Great Apes. *Proc. IXth International Congress of Anthropological and Ethnological Sciences.* The Hague: Mouton & Co.

———. 1975. Communication with Chimpanzees. In: Eibl-Eibesfeldt, I., and Kurth, G. (Eds.), *Hominisation und Verhalten.* Stuttgart: Gustav Fischer.

FOUTS, R. S., CHOWN, W., AND GOODWIN, L. Translation from vocal English to American Sign Language in a chimpanzee (Pan). *J. Learn. Linguistics* (in press).

FOUTS, R. S., AND RIGBY, R. L. Man-chimpanzee Communication.

In: Sebeok, T. A. (Ed.), *How Animals Communicate*. Bloomington: Indiana Univ. Press (in press).
FRAENKEL, G. S., AND GUNN, D. L. 1961. *The Orientation of Animals;
Kineses, Taxes and Compass Reactions*. New York: Dover (reprint of
original 1940 edition).
FRISCH, K. VON. 1923. Über die "Sprache" der Bienen. *Zool. Jahrb. Abt.
allg. Zool. Physiol. Tiere*. 40: 1–186.
———. 1946. Die Tänze der Bienen. *Österr. Zool. Z.* 1: 1–48.
———. 1950. Die Sonne als Kompass im Leben der Bienen. *Experientia
(Basel)* 6: 210–221.
———. 1967. *The Dance Language and Orientation of Bees*. (Translation by L. Chadwick.) Cambridge: Harvard Univ. Press.
———. 1972. *Bees, their Vision, Chemical Senses and Language*. 2nd
ed. Ithaca: Cornell Univ. Press.
———. 1974. Decoding the language of the bee. *Science* 185: 663–668.
GARDNER, B. T., AND GARDNER, R. A. 1969. Teaching sign language to a
chimpanzee. *Science* 165: 664–672.
———. 1971. Two-way Communication with an Infant Chimpanzee. In:
Schrier, A. M., and Stollnitz, F. (Eds.), *Behavior of Non-Human
Primates*. Vol. IV. New York: Academic Press, Chapter 3.
———. 1975. Evidence for sentence constituents in the early utterances
of child and chimpanzee. *J. Exp. Psychol.* 104: 244–267. Washoe used
a vocabulary of at least 132 signs, including many that functioned as
equivalents for nouns, proper and common, modifiers, markers such as
"time to," verbs, and locatives.
GAZZANIGA, M. S. 1975. Brain Mechanisms and Behavior. In: Gazzaniga, M. S., and Blakemore, C. (Eds.), *Handbook of Psychobiology*.
New York: Academic Press, Chapter 19.
GILL, T. V., AND RUMBAUGH, D. M. 1974. Mastery of naming skills by a
chimpanzee. *J. Hum. Evol.* 3: 483–492.
GLUCKSBERG, S., AND DANKS, J. H. 1975. *Experimental Psycholinguis-
tics; An Introduction*. Hillsdale, N.J.: Erlbaum. This book exemplifies
the limitations of perspective which I feel should be reexamined and
probably be relaxed. " . . . the dance of the honey bee shares none of
the properties of human language systems."
GOLDSTEIN, K. 1957. The Nature of Language. In: Anshen, R. V. (Ed.),
Language: An Enquiry into its Meaning and Function. New York:
Harper, Chapter 2.
GOODALL, J. VAN LAWICK. 1968. Behavior of free-living chimpanzees of
the Gombe Stream area. *Anim. Behav. Monogr.* 1: 165–311.
———. 1971. *In the Shadow of Man*. Boston: Houghton Mifflin.
———. 1975. The Behaviour of the Chimpanzee. In: Kurth, G., and

Eibl-Eibesfeldt, I. (Eds.), *Hominisation und Verhalten*. Stuttgart: Gustav Fischer.

GOULD, J. L. 1974. Honey bee communication: Misdirection of recruits by foragers with covered ocelli. *Nature* 252: 300–301.

———. 1975a. Honey bee communication: The dance-language controversy. *Science* 189: 685–693.

———. 1975b. Communication of distance information by honey bees. *J. Comp. Physiol.* 104: 161–173.

———. 1976. The dance-language controversy. *Q. Rev. Biol.* (in press).

GOULD, J. L., HENERY, J., AND MACLEOD, M. C. 1970. Communication of direction by the honey bee. *Science* 169: 544–554.

GOULD, S. J. 1975. Man and other animals. *Nat. Hist.* 84(7): 24–30.

GRAMZA, A. F. 1967. Responses of brooding nighthawks to a disturbance stimulus. *Auk* 84: 72–86.

GREGG, L. W. 1974. *Knowledge and Cognition*. Potomac, Md.: Erlbaum.

GRICE, H. P. 1957. Meaning. *Philos. Rev.* 66: 377–388.

GRIFFIN, D. R. 1946. Supersonic cries of bats. *Nature* 158: 46–48.

———. 1950. Measurements of the ultrasonic cries of bats. *J. Acoust. Soc. Am.* 22: 247–255.

———. 1953. Bat sounds under natural conditions, with evidence for the echolocation of insect prey. *J. Exp. Zool.* 123: 435–466.

———. 1958. *Listening in the Dark*. New Haven: Yale Univ. Press. (Reprinted 1974 by Dover Publications, N.Y.).

———. 1973. Echolocation. In: Møller, A. R. (Ed.), *Basic Mechanisms in Hearing*. New York: Academic Press.

———. Expanding Horizons in Animal Communication. In: Sebeok, T. A. (Ed.), *How Animals Communicate*. Bloomington: Indiana Univ. Press (in press).

GRIFFIN, D. R., AND SUTHERS, R. A. 1970. Sensitivity of echolocation in cave swiftlets. *Biol. Bull.* 139: 495–501.

GRIFFIN, D. R., WEBSTER, F. A., AND MICHAEL, C. 1960. The echolocation of flying insects by bats. *Anim. Behav.* 8: 141–154.

HAMPSHIRE, S. 1959. *Thought and Action*. London: Chatto and Windus.

HARTRIDGE, H. 1920. The avoidance of objects by bats in their flight. *J. Physiol.* (*Lond.*) 54: 54–57.

HARTSHORNE, C. 1973. *Born to Sing. An Interpretation and World Survey of Bird Song*. Bloomington: Indiana Univ. Press.

HATTIANGADI, J. N. 1973. Mind and the origin of language. *Philos. Forum* 14: 81–98.

HEALY, A. F. 1971. Can chimpanzees learn a phonemic language? *J. Psycholinguist. Res.* 2: 167–170.

HEBB, D. O. 1974. What psychology is about. *Am. Psychol.* 29: 71–79.
HEWES, G. W. 1974. Gesture language in culture contact. *Sign Language Studies* 1: 1–34.
———. 1975. *Language Origins, A Bibliography*. 2nd. ed. The Hague: Mouton. Vols. 1 and 2. A lengthy bibliography with many brief comments concerning the enormous literature on theories concerning the origin of human language.
HINDE, R. A. 1970. *Animal Behaviour: A Synthesis of Ethology and Comparative Psychology*. New York: McGraw-Hill.
———. (Ed.) 1972. *Non-verbal Communication*. London: Cambridge Univ. Press.
HOCKETT, C. F. 1958. *A Course in Modern Linguistics*. New York: Macmillan.
HOCKETT, C. F., AND ALTMANN, S. A. 1968. A Note on Design Features. In: Sebeok, T. A. (Ed.), *Animal Communication*. Bloomington: Indiana Univ. Press, Chapter 5.
HOLLOWAY, R. L. 1974. Review of Jerison, H. J. *Evolution of the Brain and Intelligence. Science* 184: 677–679.
HOPKINS, C. D. 1974. Electrical communication in fish. *Am. Sci.* 62: 426–437.
HUBBARD, J. I. 1975. *The Biological Basis of Mental Activity*. Reading, Mass: Addison-Wesley. A short and moderately elementary textbook with an innovative approach which leads the author to include a final chapter entitled "Brain and Mind." The conclusion reached is that no separate mental entities are necessary and that mental phenomena result directly from the activities of nervous systems.
HUTCHINSON, G. E. 1976. Man talking or thinking. *Am. Sci.* 64: 22–27.
IRWIN, F. W. 1971. *Intentional Behavior and Motivation*. Philadelphia: Lippincott.
JENNINGS, H. S. 1906. *Behavior of Lower Organisms*. New York: Columbia Univ. Press. (Reprinted with preface by D. D. Jensen, Bloomington: Indiana Univ. Press, 1962.)
———. 1910. Diverse ideals and divergent conclusions in the study of behavior in lower organisms. *Am. J. Psychol.* 21: 349–370.
———. 1933. *The Universe and Life*. New Haven: Yale Univ. Press.
KALMIJN, A. J. 1971. The electric sense of sharks and rays. *J. Exp. Biol.* 55: 371–383.
———. 1974. The detection of electric fields from inanimate and animate sources other than electric organs. In: Fessard, A. (Ed.), *Electroreceptors and Other Specialized Receptors in Lower Vertebrates*. New York: Springer, Chapter 5.
KEETON, W. 1974. The Orientational and Navigational Basis of Homing

in Birds. In: Lehrman, D., Hinde, R., and Shaw, E. (Eds.), *Advances in the Study of Behavior*. Vol. 5. New York: Academic Press.

KELLOGG, W. N. 1961. *Porpoises and Sonar*. Chicago: Chicago Univ. Press.

KENNY, A. J. P., LONGUET-HIGGINS, H. C., LUCAS, J. R., AND WADDINGTON, C. H. 1972. *The Nature of Mind*. Edinburgh: Edinburgh Univ. Press.

———. 1973. *The Development of Mind*. Edinburgh: Edinburgh Univ. Press.

KIMBLE, G. A., AND PERLMUTER, L. C. 1970. The problem of volition. *Psychol. Rev.* 77: 361–384. This paper begins with a pertinent quotation from C. S. Sherrington's *Integrative Action of the Nervous System* (New York: Scribner, 1906): "Yet it is clear, in higher animals especially so, that reflexes are under control . . . the reactions of reflex-arcs are controllable by mechanisms to whose activity consciousness is adjunct." Kimble and Perlmuter conclude ". . . that a voluntary act begins with an idea of the response to be performed . . . ." They also assume that "voluntary behavior is always learned," and they believe that "comparator mechanisms" must be present so that sensory input can be compared with the image assumed to be present internally.

KLEIN, D. B. 1970. *A History of Experimental Psychology*. New York: Basic Books.

KLIMA, E. S., AND BELLUGI, V. 1973. *Teaching Apes to Communicate*. In: Miller, G. A. (Ed.), *Communication, Language, and Meaning*. New York: Basic Books, Chapter 9.

KLOPFER, P. H., AND HAILMAN, J. P. 1967. *An Introduction to Animal Behavior*. Englewood Cliffs, N.J.: Prentice-Hall.

KÖHLER, W. 1925. *The Mentality of Apes*. London: K. Paul, Trench, Trubner. New York: Harcourt Brace.

KRAMER, G. 1959. Recent experiments on bird orientation. *Ibis* 101: 399–416.

KRAMES, L., PLINER, P., AND ALLOWAY, T. (Eds.) 1974. *Non-verbal Communication*. New York: Plenum.

KREBS, H. A. 1975. The August Krogh principle: "For many problems there is an animal on which it can be most conveniently studied." *J. Exp. Zool.* 194: 221–226.

KREITHEN, M. L., AND KEETON, W. T. 1974a. Detection of polarized light by the homing pigeon, *Columba livia*. *J. Comp. Physiol.* 89: 83–92.

———. 1974b. Attempts to condition homing pigeons to magnetic stimuli. *J. Comp. Physiol.* 89: 83–92.

LANGER, S. K. 1942. *Philosophy in a New Key.* New York: Pelican Books.

———. 1962. *Philosophical Sketches.* Baltimore: Johns Hopkins Press. " . . . language is symbolic, when no animal utterance shows any tendency that way. The biological factors that caused this great shift in the vocal function were, I believe, the development of visual imagery in the humanoid brain, and the part it came to play in a highly exciting, elating experience, the festal dance." This was written only a few years before Goodall (1968, 1971) described what appeared to be highly excited "rain dances" in chimpanzees. Langer continued: "As I remarked before, images are more prone than anything else we know to become symbols. . . . In animals typically, every stimulation that takes effect at all is spent in some overt act...."

———. 1967. *Mind: An Essay on Human Feeling.* Vol. I. Baltimore: Johns Hopkins Press.

———. 1972. *Mind: An Essay on Human Feeling.* Vol. II. Baltimore: Johns Hopkins Press. "A genuine symbol is, above all, an instrument of conception, and cannot be said to exist short of meeting that requirement; that means that an ape thinking symbolically could think of an act he had no intention or occasion to perform, and envisage things entirely remote from his real situation. . . . Symbolism is the mark of humanity." These views may have been expressed before the full impact of the Gardners' breakthrough concerning chimpanzee communication had been felt, and before the experiments of Menzel and Halperin. But it is now clear that some animals communicate complex messages so closely atuned to the nuances of the social situation that great caution is called for in reaching such definite conclusions as those expressed by Langer.

LASHLEY, K. S. 1923. The behavioristic interpretation of consciousness. *Psychol. Rev.* 30: 237–272 and 329–353.

———. 1949. Persistent problems in the evolution of mind. *Q. Rev. Biol.* 24: 28–42. Here Lashley restates the basic behavioristic position: "The questions of where mind or consciousness enters in the phylogenetic scale and of the nature of conscious experience as distinct from physiological processes are pseudo-problems, arising from misconceptions of the nature of the data revealed by introspection. A comparative study of the behavior of animals is a comparative study of mind, by any meaningful definition of the term."

———. 1958. Cerebral organization and behavior. *Proc. Assoc. Res. Nerv. Ment. Dis.* 36: 1–18. (Reprinted in Beach, F. A., Hebb, D. O., Morgan, C. T., and Nissen, H. W., *The Neuropsychology of Lashley.* New York: McGraw-Hill, 1960.)

LENNEBERG, E. H. 1971. Of language knowledge, apes, and brains. *J. Psycholinguist. Res.* 1: 1–29.

LINDAUER, M. 1955. Schwarmbienen auf Wohnungssuch. *Z. vgl. Physiol.* 37: 263–324. "Veilleicht ist dies mit ein entscheidender Punkt, um eine Einigung zustande zu bringen, dass die Spurbienen nicht hartnäckig bei ihrem ersten Urteil verbleiben, sondern nach kürzerer oder längerer Zeit verstummen und den weiteren Entscheid den Neulingen überlassen. 'Jetzt sollen die ihr Urteil abgehen.' . . . Das ist mit Absicht anthropomorphistisch augsgedrükt. Selbstverständlich wird eine Spurbeine nicht bewusst ein Urteil bilden, wie es unserem menschlichen Begriff entspricht." Perhaps this is to be brought together with a crucial point, namely that the scout bees do not cling stubbornly to their first opinion, but, after a shorter or longer time, fall silent and yield further decision to the newcomers: "Now you give your opinion." This is intentionally expressed in anthropomorphic terms. But it is self-evident that a scout bee does not consciously form an opinion corresponding to our human concepts.

———. 1971a. *Communication among Social Bees* (revised edition). Cambridge: Harvard Univ. Press.

———. 1971b. The functional significance of the honey bee waggle dance. *Am. Nat.* 105: 89–96.

LINDAUER, M., AND MARTIN, H. 1972. Magnetic Effect on Dancing Bees. In: Galler, S. R., Schmidt-Koenig, K., Jacobs, G. J., and Belleville, R. E. (Eds.), *Animal Orientation and Navigation.* Washington, D.C.: U.S. Government Printing Office.

LINDEN, E. 1974. *Apes, Men, and Language.* New York: Dutton.

LISSMANN, H. W. 1958. On the function and evolution of electric organs in fish. *J. Exp. Biol.* 35: 156–191.

LISSMANN, H. W., AND MACHIN, K. E. 1958. The mechanism of object location in *Gymnarchus niloticus* and similar fish. *J. Exp. Biol.* 35: 415–486.

LLOYD, J. E. 1966. *Studies on the Fish Communication System in Photinus Fireflies.* Ann Arbor, Mich. Misc. Publ. No. 130, Museum of Zoology, Univ. of Michigan.

———. 1975. Aggressive mimicry in *Photurus* fireflies: signal repertoires by femmes fatales. *Science* 187: 452–453.

LOEB, J. 1900. *Comparative Physiology of the Brain and Comparative Psychology.* New York: Putnam.

———. 1912. *The Mechanistic Conception of Life.* Chicago: Univ. of Chicago Press. (Reprinted with preface by D. Fleming, Cambridge, Mass: Harvard Univ. Press, 1964.)

———. 1916. *The Organism as a Whole, from the Physicochemical Viewpoint.* New York: Putnam.

LORENZ, K. 1958. *Methods of Approach to the Problems of Behavior.* Harvey Lectures 1958–59. New York: Academic Press. (Reprinted in *Studies in Animal and Human Behavior.* Vol. II. Cambridge, Mass: Harvard Univ. Press, 1971.)

———. 1963. Haben Tiere ein subjectives Erleben? *Jahr. Techn. Hochs. Munchen.* (English translation reprinted in *Studies in Animal and Human Behavior.* Vol. II. Cambridge, Mass: Harvard Univ. Press, 1971.)

———. 1969. Innate Bases of Learning. In: Pribram, K. H. (Ed.), *On The Biology of Learning.* New York: Harcourt, Brace and World. In a wide-ranging discussion of ethology and its implications, Lorenz reviews evidence that cultural transmission occurs in many examples of social behavior and communication among animals.

MALCOLM, N. 1973. *Thoughtless Brutes.* Proceedings and Addresses of American Philosophical Association, 46: 5–20. Malcolm quotes Descartes as stating only that "it could not be *proved* either that animals do or that they do not have thoughts 'hidden in their bodies.' . . . But the idea that we cannot determine whether dogs have thoughts *in* them is a dreadful confusion. . . . The relevant question is whether they *express* thoughts. I think the answer is clearly in the negative. . . The possession of language makes the whole difference. . . . An undertaking of trying to find out whether the dog did or didn't have that thought is not anything we understand.

"Descartes' notion was that speech is the only 'sign' of the presence of thought . . . that just conceivably animals may have thoughts of which they give no sign. This implies a looseness of connection between thought and the linguistic expression of thought, that is deeply disturbing. . . . The relationship between language and thought must be . . . so close that it is really senseless to conjecture that people may *not* have thoughts, and also really senseless to conjecture that animals *may* have thoughts."

MARITAIN, J. 1957. Language and the Theory of Sign. In: Anshen, R. N. (Ed.), *Language: An Inquiry into its Meaning and Function.* New York: Harper & Row, Chapter 5.

MARLER, P. 1968. Visual Systems. In: Sebeok, T. A. (Ed.), *Animal Communication.* Bloomington: Indiana Univ. Press, Chapter 7.

———. 1969. Animals and Man: Communication and its Development. In: Roslansky, J. D. (Ed.), *Communication.* Amsterdam: North-Holland, pp. 25–61.

———. 1974. Animal Communication. In: Krames, L., Pliner, P., and Alloway, T. (Eds.), *Nonverbal Communication.* New York: Plenum, Chapter 2.

———. Affective and symbolic meaning: some zoosemiotic speculations (in preparation).

MARLER, P., AND HAMILTON, W. H. 1966. *Mechanisms of Animal Behaviour.* New York: Wiley.

MARSHALL, A. J. 1954. *Bower-birds, Their Displays and Breeding Cycles.* London: Oxford Univ. Press.

MARTIN, H., AND LINDAUER, M. 1973. Orientierung im Erdmagnetfeld. *Fortschr. Zool.* 21: 211–228.

MATTHEWS, G. V. T. 1968. *Bird Navigation.* 2nd ed. London: Cambridge Univ. Press.

MCGREW, W. C. 1974. Tool use by wild chimpanzees in feeding upon driver ants. *J. Hum. Evol.* 3: 501–508.

MCGUIGAN, F. J., AND SCHOONOVER, R. A. 1973. *The Psychophysiology of Thinking: Studies of Covert Processes.* New York: Academic Press.

MCMULLAN, E. M. 1969. Man's Effort to Understand the Universe. In: Roslansky, J. D. (Ed.), *The Uniqueness of Man.* Amsterdam: North-Holland. McMullan takes it for granted that at least some mental experiences occur in animals, but he argues against extension to nonhuman animals of subjectivity of consciousness awareness. He objects to what he considers two extreme views, one labeled "behaviorist" and the other "ethologist," which he defines by hypothetical quotations: Behaviorist: "there is no more to thinking than the overt behavior elicited and the accompanying neurological changes"; and Ethologist: "the honey-bee's use of language indicates a symbolic intelligence of the sort at work in human language, only at a simpler level." McMullan asserts that both extreme views "tend to smooth out the differences between organisms." He rejects the dances of bees as a language because "First, they are species-specific, inherited not learnt. Their use is instinctive, not reflective. Honey-bees of one species will not be able to 'follow' the language of another species, nor can they learn it."

MENZEL, E. W., JR. 1974. A Group of Young Chimpanzees in a One-acre Field. In: Schrier, A. M., and Stollnitz, F. (Eds.), *Behavior of Nonhuman Primates.* Vol. 5. New York: Academic Press.

MENZEL, E. W., JR., AND HALPERIN, S. 1975. Purposive behavior as a basis for objective communication between chimpanzees. *Science* 189: 652–654.

MICHENER, C. D. 1974. *The Social Behavior of the Bees.* Cambridge, Mass: Harvard Univ. Press.

MILLER, G. A. 1962. *Psychology, the Science of Mental Life.* New York: Harper & Row.

———. 1967. *The Psychology of Communication.* New York: Basic Books.

MILLER, G. A., GALANTER, E., AND PRIBRAM, K. H. 1960. *Plans and the Structure of Behavior.* New York: Holt, Rinehart and Winston, Chapter 11.

MITTELSTAEDT, H. 1972. Kybernetik der Schwereorientierung. *Verh. Dtsch. Zool. Ges.* (1972) 185–200.

MOEHRES, F. P., AND OETTINGEN-SPIELBERG, T. 1949. Versuche über die Nahorientierung und Heimfindevermögen der Fledermäuse. *Verh. Dtsch. Zool. Ges.* (1949) 248–252.

MONOD, J. L. 1975. On Molecular Theory of Evolution. In: Harre, R. (Ed.), *Problems of Scientific Revolution, Progress and Obstacles to Progress in the Sciences.* London: Oxford Univ. Press. ". . . man is endowed with a completely unique capacity, which no other species shares, namely language. . . . There is nothing argumentative for instance, in animal communication" (page 23).

MORGAN, C. L. 1894. *An Introduction to Comparative Psychology.* London: Scott.

MORRIS, C. 1946. *Signs, Language, and Behavior.* Englewood Cliffs, N.J.: Prentice-Hall. (Reprinted 1964. New York: Braziller). Morris uses as an example of the difference between his usage of sign and symbol a situation in which a dog has been conditioned to respond to the sound of a buzzer by going to some place other than the buzzer to obtain food. ". . . the advantage of such symbols is found in the fact that they occur in the absence of signals produced by the environment; an action or state of the interpreter itself becomes (or produces) a sign guiding behavior with respect to the environment. So if a symbol operates in the dog's behavior, the symbol could take the place in the control of behavior which the buzzer formerly exercised: hunger cramps for instance might themselves come to be a sign (that is, a symbol) of food at the customary place." One may well decide that such semantic exercises as these are of dubious value in dealing with complex communication behavior such as the waggle dances of honeybees, which were unknown when Morris developed his definitions. Conversely, if it seems more reasonable to consider the waggle dance as a signal-process rather than a symbol-process in Morris's terms, this will have no important bearing on the principal considerations under discussion.

MOWRER, O. H. 1960a. *Learning Theory and Behavior.* New York: Wiley.

———. 1960b. *Learning Theory and Symbolic Processes.* New York: Wiley.

NANCE, J. 1975. *The Gentle Tasaday, A Stone Age People in the Philippine Rain Forest.* New York: Harcourt Brace Jovanovich. The discussion included in this charming book of first contacts with a newly discovered tribe indicates that a few words were shared by the Tasaday and the hunter who first made contact with them. The situation in which adult human beings suddenly come into contact without sharing any words whatsoever seems at the present time to be so rare that it has received very little attention from anthropologists or linguists, except for Hewes (1974, 1975).

NEBES, R. D., AND SPERRY, R. W. 1971. Hemispheric deconnection syndrome with cerebral birth injury in the dominant arm area. *Neuropsychologia* 9: 247–259.

NEISSER, U. 1967. *Cognitive Psychology*. New York: Appleton-Century-Crofts.

NORRIS, K. S. (Ed.) 1966. *Whales, Dolphins, and Porpoises*. Berkeley: Univ. of California Press.

PFUNGST, O. 1911. *Clever Hans, the Horse of Von Osten*. (With preface of J. R. Angell and introduction by C. Stumpf.) New York: Holt, Rinehart and Winston. (Reprinted 1965 in English translation by C. L. Rahn and introduction by R. Rosenthal.)

PIAGET, J. 1971. *Biology and Knowledge: An Essay on the Relations between Organic Regulations and Cognitive Processes*. Chicago: Univ. of Chicago Press. (Translation by B. Walsh of *Essai sur les Relations entre les Regulations Organiques et les Processus Cognitifs*. Paris: Gallimard, 1967.)

POLLIO, H. R. 1974. *The Psychology of Symbolic Activity*. Reading, Mass.: Addison-Wesley. A review of honeybee communication which the author does not accept as a language.

POLTEN, E. P. 1973. *Critique of the Psycho-Physical Identity Theory*. The Hague: Mouton. This book presents counterarguments to those of Feigl. Animals are denied free will (page 163) even though the author admits on page 230 that "we cannot be sure of the exact type of internal consciousness of various animals." On page 160, he states: "It does seem indeed true that animals are not *aware* what they are experiencing or doing, cannot conceptualize or think." On page 129 Polten quotes Hegel: "The only mere physicists are the animals: they alone do not think: while man is a thinking being and a born metaphysician."

POPPER, K. R. 1972. *Objective Knowledge. An Evolutionary Approach*. London: Oxford Univ. Press. On page 74 Popper tells us " . . . Animals, although capable of feelings, sensations, memory, and thus of consciousness, do not possess the full consciousness of self which is one of the results of human language. . . . " On page 122 he continues, "The most important functions or dimensions of the human language (which animal languages do not possess) are the descriptive and argumentative functions."

————. 1974. Scientific Reduction and the Essential Incompleteness of all Science. In: Ayala, F. J., and Dobzhansky, T. (Eds.), *Studies in the Philosophy of Biology, Reduction and Related Problems*. Berkeley: Univ. of California Press, Chapter 16. "Though this behaviorist philosophy is quite fashionable at present, a theory of the nonexistence of consciousness cannot be taken anymore seriously, I suggest, than a theory of the nonexistence of matter."

PREMACK, D. 1971. On the Assessment of Language Competence and the Chimpanzee. In: Schrier, A. M., and Stollnitz, F. (Eds.), *Behavior of Non-Human Primates*. Vol. 4. New York: Academic Press, Chapter 4.

————. 1975. Symbols Inside and Outside of Language. In: Kavanagh, J. F., and Cutting, J. E. (Eds.), *The Role of Speech in Language*. Cambridge, Mass.: M. I. T. Press.

PRIBRAM, K. H. 1971. *Languages of the Brain*. Englewood Cliffs, N.J.: Prentice-Hall.

PRICE, H. H. 1938. Our evidence for the existence of other minds. *Philosophy* 13: 425–456.

PYLES, T. 1971. *The Origins and Development of the English Language*. 2nd ed. New York: Harcourt Brace Jovanovich.

RAZRAN, G. 1971. *Mind in Evolution, an East-West Synthesis of Learned Behavior and Cognition*. Boston: Houghton-Mifflin.

RENSCH, B. 1971. *Biophilosophy*. (Translation of 1968 German edition by C. A. M. Sym.) New York: Columbia Univ. Press. Rensch accepts the likelihood that animals have mental experiences and consciousness, but feels that animal concepts differ from human concepts in not being combined with words.

ROBINSON, D. N. 1973. *The Enlightened Machine: An Analytical Introduction to Neurophysiology*. Encino, California: Dickenson. In the final chapter of this semipopular survey of neurophysiology, the author begins to define language in terms of a parrot's "Polly wanna cracker," honeybee dances, and clouds that "inform us of impending rain, and when they take on a certain texture and mass, they tell us to find shelter. We reject the clouds' communication as language because, we insist, the clouds do not *know* what they are telling us; they do not *intend* to inform us. . . . There is one thing that babies, puppies, parrots, and clouds have in common: *They cannot lie*."

ROEDER, K. 1970. Episodes in insect brains. *Am. Sci.* 58: 378–389.

ROEDER, K., AND TREAT, A. 1957. Ultrasonic reception by the tympanic organ of noctuid moths. *J. Exp. Zool.* 134: 127–157.

ROHLES, F. H., AND DEVINE, J. V. 1966. Chimpanzee performance on a problem involving the concept of middleness. *Anim. Behav.* 14: 159–162.

———. 1967. Further studies of the middleness concept with the chimpanzee. *Anim. Behav.* 15: 107–112.

RUMBAUGH, D. M., VON GLASERFELD, E., WARNER, H., PISANI, P., AND GILL, T. V. 1974. Lana (chimpanzee) learning language: A progress report. *Brain Language* 1: 205–212.

SARLES, H. 1975. Language and Communication—II: The View from '74. In: Pliner, P., Krames, L., and Alloway, T. (Eds.), *Nonverbal Communication of Aggression.* New York: Plenum.

SAUER, E. G. F. 1957. Die Sternenorientierung nächtlich ziehender Grasmücken (*Sylvia atricapilla, borin* und *curruca*). *Z. Tierpsychol.* 14: 29–70.

SCHMIDT-KOENIG, K. 1965. Current Problems in Bird Orientation. In: Hinde, R., Lehrman, D., and Shaw, E. (Eds.), *Advances in the Study of Behavior.* Vol. 1. New York: Academic Press.

SCHULTZ, D. 1975. *A History of Modern Psychology.* 2nd ed. New York: Academic Press.

SCRIVEN, M. 1963. The Mechanical Concept of Mind. In: Sayre, K. M., and Crosson, F. J. (Eds.), *The Modeling of Mind, Computers and Intelligence.* Notre Dame, Ind.: Notre Dame Press. "Consciousness is, it now seems to me, automatically guaranteed by the capacity to categorize and discuss one's own reactions, beliefs, etc." (p. 254).

SEBEOK, T. A. 1968. (Ed.). *Animal Communication.* Bloomington: Indiana Univ. Press.

———. 1972. (Ed.). *Perspectives in Zoosemantics.* The Hague: Mouton.

SEBEOK, T. A., AND RAMSAY, A. 1969. (Eds.). *Approaches to Animal Communication.* The Hague: Mouton.

SIMMONS, J. A., HOWELL, D. J., AND SUGA, N. 1975. The information content of bat sonar echoes. *Am. Sci.* 63: 204–215.

SIMPSON, G. G. 1964. *Biology and Man.* New York: Harcourt, Brace & World, Chapter 8.

SKINNER, B. F. 1957. *Verbal Behavior.* New York: Appleton-Century-Crofts.

SKUTCH, A. F. 1976. *Parent Birds and their Young.* Austin, Texas: Univ. of Texas Press.

"SMITH, ADAM." 1972. *Supermoney.* New York: Popular Library.

———. 1975. *Powers of Mind.* New York: Random House.

SMITH, K. 1969. *Behavior and Conscious Experience, A Conceptual Analysis.* Athens: Ohio Univ. Press. Smith advocates a form of "physical monism" according to which " . . . awareness itself is a proper part of nature." Conscious experience is defined as "an internal event to which an arbitrary response can be attached directly, by the process of following that response, when it occurs to that event, with a prespecified set of circumstances." Smith believes, despite the behaviorists, that psychology is "the science of the nature, causes, and effects of conscious experience—conscious experience being specified always in terms of the stimulus situations instigating it." He deplores "the somewhat remarkable insistence of modern psychology on avoiding the whole topic of conscious experience..." and he advocates instead what he calls "introspectional behaviorism."

SMITH, W. J. 1968. Message-meaning Analysis. In: Sebeok, T. A. (Ed.), *Animal Communication*. Bloomington: Indiana Univ. Press.

———. 1969. Messages of vertebrate communication. *Science* 165: 145–150.

———. 1975. Communication, animal. *Encyclopedia Britannica, Macropedia*. Vol. 4, pp. 1010–1019. Chicago: Encyclopedia Britannica.

———. *The Behavior of Communication: An Ethological Approach*. Cambridge, Mass.: Harvard Univ. Press (in press).

SPERRY, R. W. 1969. A modified concept of consciousness. *Psychol. Rev.* 76: 532–536. (Also discussion with D. Bindra, ibid. 77:581–590, 1970.)

———. 1973. Lateral Specialization of Cerebral Function in the Surgically Separated Hemispheres. In: McGuigan, F. J., and Schoonover, R. A. (Eds.), *The Psychophysiology of Thinking, Studies of Covert Processes*. New York: Academic Press. This paper and the group discussion published with it demonstrate that both hemispheres of the human cerebral cortex are capable of complex mental processing, and that while normally the dominant hemisphere (usually the left) handles linguistic processes, the division of labor is variable among individuals and probably not sharp and total.

STENHOUSE, B. 1973. *The Evolution of Intelligence*. London: George Allen and Unwin.

STEVENSON, J. C. 1969. Song as a Reinforcer. In: Hinde, R. A. (Ed.), *Bird Vocalizations*. London: Cambridge Univ. Press.

STOUT, J. F., AND BRASS, M. E. 1969. Aggressive communication by *Larus glaucescens*. Part II. Visual communication. *Behaviour* 34: 42–54.

STOUT, J. F., WILCOX, C. R., AND CREITZ, L. E. 1969. Aggressive communication by *Larus glaucescens*. Part I. Sound communication. *Behaviour* 34: 29–41.

TAVOLGA, W. N. 1974. Application of the Concept of Levels of Organization to the Study of Animal Communication. In: Krames, L., Pliner, P., and Alloway, T. (Eds.), *Nonverbal Communication*. New York: Plenum.

Tax, S.. and Callender, C. (Eds.). 1960. Evolution after Darwin. *Issues in Evolution*. Vol. III. Chicago: Univ. of Chicago Press.

Taylor, J. G. 1962. *The Behavioral Basis of Perception*. New Haven: Yale Univ. Press. In his final chapter, Taylor concludes that there is no need to postulate any special essence to explain consciousness. "Thus the revolt of the early behaviorists against the 'mentalist' assumptions of classical psychology has led, through the assiduous investigation of the laws of behavior, to a new assertion of the scientific respectability of the once despised phenomena of consciousness."

Teng, E. L., and Sperry, R. W. 1973. Interhemispheric interaction during simultaneous bilateral presentation of letters or digits in commisurotomized patients. *Neuropsychologia* 11: 131–140.

Terwilliger, R. F. 1968. *Meaning and Mind, a Study of the Psychology of Language*. New York: Oxford Univ. Press. This book is an example of the viewpoint that " . . . the nature of one's language *determines* his entire way of life, including his thinking and all other forms of mental activity."

Thass-Thienemann, T. 1968. *Symbolic Behavior*. New York: Washington Square Press.

Thomas, L. 1974. *The Lives of a Cell, Notes of a Biology Watcher*. New York: Viking.

Thorpe, W. H. 1972. In: Hinde, R. A. (Ed.), *Nonverbal Communication*. London: Cambridge Univ. Press, Chapters 2, 5, and 6.

———. 1974a. *Animal Nature and Human Nature*. Garden City, N.Y.: Doubleday.

———. 1974b. Reductionism in Biology. In: Ayala, F. J., and Dobzhansky, T. (Eds.), *Studies in the Philosophy of Biology, Reduction and Related Problems*. Berkeley: Univ. of California Press, Chapter 8.

Tinbergen, N. 1951. *The Study of Instinct*. London: Oxford Univ. Press.

Tolman, E. C. 1932. *Purposive Behavior in Animals and Men*. New York: Century.

———. 1948. Cognitive maps in rats and men. *Psychol. Rev.* 55: 189–208.

Walcott, C., and Green, R. P. 1974. Orientation of homing pigeons altered by a change in the direction of an applied magnetic field. *Science* 184: 180–182.

Watson, J. B. 1929. *Psychology from the Standpoint of a Behaviorist*. Philadelphia: Lippincott.

Weiner, B. 1972. *Theories of Motivation, from Mechanism to Cognition*. Chicago: Markham Publishing Co. In a section on Cognitive Psychology of Infrahumans, Weiner states that "it is reasonable to presume that lower organisms also have cognitions. In infrahumans, however, cognitions are not directly assessible via introspective reports. But they may be inferred from behavior. . . . Morgan's Canon should be considered a guiding statement, rather than an invariant law."

WEIS, D. D. 1975. Professor Malcolm on animal intelligence. *Philos. Rev.* 84: 88–95.

WELLS, P. H. 1973. Honey Bees. In: Corning, W. C., Dyal, J. A., and Willows, A. O. D. (Eds.), *Invertebrate Learning.* Vol. 2. New York: Plenum, Chapter 9. Wells dismisses completely the evidence that bee dances serve any communicatory function.

WELLS, P. H., AND WENNER, A. M. 1973. Do honey bees have a language? *Nature (Lond.)* 241: 171–175.

WENNER, A. M. 1962. Sound production during the waggle dance of the honey bee. *Anim. Behav.* 10: 79–95.

———. 1971. *The Bee Language Controversy.* Boulder, Colo.: Educational Programs Improvement Corp.

———. 1974. Information Transfer in Honey Bees: A Population Approach. In: Krames, L., Pliner, P., and Alloway, T. (Eds.), *Nonverbal Communication.* New York: Plenum.

WESTBY, G. W. M. 1974. Assessment of the signal value of certain discharge patterns in the electric fish, *Gymnotus carapo,* by means of playback. *J. Comp. Physiol.* 92: 327–341.

WHITEHEAD, A. N. 1938. *Modes of Thought.* New York: Macmillan.

WILCOX, R. S. 1972. Communication by surface waves: Mating behavior of a water strider (Gerridae). *J. Comp. Physiol.* 80: 225–266.

WILSON, E. O. 1971. *The Insect Societies.* Cambridge, Mass: Harvard Univ. Press.

———. 1975. *Sociobiology, the New Synthesis.* Cambridge, Mass: Harvard Univ. Press.

WILTSCHKO, W. 1974. Der Magnetkompass der Gartengrassmücke *(Sylvia borin). J. Ornithol.* 115: 1–7.

———. 1975. The interaction of stars and magnetic field in the orientation system of night migrating birds. *Z. Tierpsychol.* 37: 337–355.

WITTGENSTEIN, L. 1953. *Philosophical Investigations.* 3rd ed. (Translated by Anscombe, G. E. M.). New York: Macmillan.

SUBJECT INDEX

A

affect 58, 98
affection 6
altruistic behavior vii
American Sign Language, learned by chimpanzees 17
Andrea Doria bats 13
animal communication
  conscious thought and 39
  versatility of 15 ff.
animal surrogates 95
anthropologists
  new language and 88
anthropomorphism 68, 72, 76, 104
antibodies 69
anticipation 50, 84
ants
  "ganglion on legs" 44
  lead others to food 25
  stridulation by 16
  tandem running by 25
Apis florea 24
arenas, mating 77
astrophysics 59
attention 57
awareness 14, 39 ff., 58, 82, 101
  adaptive advantage of 81 ff., 104
  definition 5

B

bats
  Andrea Doria 13
  echolocation by 10, 11, 13
  spatial memory of 13
bees, *see* honeybees
behavior
  adaptive 85
  complexity of 52
  covert 67
  covert nonverbal 69
  covert verbal 58, 65, 84
  higher and lower 47
  prediction of 74
behavioral sciences 71
behaviorism 25, 60, 63, 71, 73
  definition 39
  objections to 49, 57, 68, 100, 104

purposive 48
biological clocks vii, 67
birds 66
  communication by 25, 90
  displays by 76, 77
  echolocation in caves 11
  songs of 78
bony structure 69
bowers, of bower birds 76
brains, minds and 7, 53, 65

C

call notes of bats 10
caloric value of tissues 73
Camponotus 25
Cardiocondyla venustula 25
Cartesian machines 67
cats, communication with 89
cellular respiration 53
cerebral dominance 51
chemical bonds 59
chemical communication 16
chimpanzees, *see also* Washoe
  communication
    by artificial symbols 17
    covert 66
    gestural 16, 69, 88, 97, 101
  concepts in 43, 83
  denied true minds 42
  facial expressions of 19
  impersonation of 95
  lying by 19
  Sarah 58, 84
  sign language of 50, 103
  speech difficulty 16
  tool-using by 54
choice 58
Clever Hans 26, 72
clocks, biological vii, 67
cognitive ethology vii, 102, 105
cognitive maps 13
cognitive processes in animals 48, 49, 97
colors, subjectivity of 6
combustion, heat of 73
communication, animal 15, 16, 53
  bird 25, 90

channels 88
design features of 34 ff.
electric fish 91
firefly 16
nonverbal 65, 88
spider 16
symbolic 26, 90, 103
two-way with animals 89, 90, 105
with environment 8
communication behavior
covert 67
reinforcement by 91
variability of 45
comparative linguistics 95 ff.
computers and mental experiences 6, 42
programs of 53
concept 58
of chimpanzees 83
denied to animals 43
conflict behavior 41
conscious awareness 52
conscious intentions 44, 48
consciousness 48, 58
in animals 14, 41, 55, 69, 101, 103
cerebral excitation and 51
defined 5, 84
language and 69
limited to language users 43
meaningless to behaviorists 5, 64
conscious thought, suggested by animal communication 39
context-dependence of animal signals 91
continuity, evolutionary 55, 98
of mental experience 68
Copernican revolution 54, 69, 99
corpus callosum 51
courtship 81
of bower birds 76
covert communication 67
nonverbal 69
verbal 58, 65, 84
crabs, fiddler 15
enlarged claws of 81
creativity of language 69
critical standards
relaxation of 71

D

dances of honeybees, see honeybee dances
Darwinian revolution 54, 69, 99
death awareness 50, 51
deception, animal 19, 45
decisions by animals 43
delayed responses by animals 67
Dendrocopos major
communication by drumming 25
design features of language 34 ff.
desires
animal 60
human 6
dialogues with animals 91
displacement 75
DNA 46, 61
dogs, communication with 89
dolphins, echolocation by 11
drives, psychological 59
dwarf honeybee 24

E

echo frequency spectra 11
echolocation
adaptation for orientation 13
by bats 10, 11
by marine mammals 11
electric fish 83
communication 16
with models 91
orientation of 11
emotions in animals 98
endogenous activity rhythms vii
environmental influences 61
estheticism of birds 77
ethology vii, 3 ff.
cognitive 102, 105
evolution 57, 99
arguments on 86
continuity of 55, 96
mental experiences and 55, 70
expectancy 49, 58
experience
conscious 51
subjective 49
experiments, need for 99

F

fears, human 6
feelings 6, 58, 60, 61
    animal 42, 98
    bird 77
fiddler crabs 15
    enlarged claws of 81
finches, Darwin's 54
firefly communication 16
fish
    electrical communication by 16
    electrical orientation by 11
    magnetic sensitivity of 12
flying phalangers, chemical communi-
        cation in 16
free will 58

G

"ganglion on legs" 44
genes 59
genetic influences 61
Gestalt 83, 90
gestures 65, 88, 89
    by chimpanzees 16
    as symbols 24
gravitation 59
gulls
    glaucous-winged 90
    herring 78
    laughing, communication of 91
    soaring of 78

H

hearing, ultrasonic, of insects 11
honeybee dances 42, 44, 53, 66, 82, 92,
        98, 103
    acoustic components of 93
    agreement via 23
    altered by other dances 23
    artificial light and 27
    and colony's needs 22
    communication by 19 ff., 92
        Gould's test of 27
        skepticism about 26
    direction signaled by 20
    food sources and 22, 24
    internal states 27
    misdirection experiments 27

    possible "finer grain" of 94
    round 20
    Schwirrlauf 24
    swarming and 22, 59
    vertical information and 94
    vigor of 93
    waggle 20
honeybees
    dwarf 24
    landmarks used by 67
    magnetic sensitivity of 12
    ocelli of 28
    polarized-light orientation 11
hope
    denied to animals 50
    human 6
horizontal surfaces, honeybee
        dances on 21
hormones 59
horses, *see* Clever Hans
human uniqueness 54, 63, 103
humanists, scientists and 41
hunger 6, 78
hypotheses, testable 99

I

images
    internal 14, 58, 84
    mental 57, 84
"impersonation" of animals 95
individual recognition viii, 15
    by birds 91
inhibition of response 67
injury-feigning 75
innate ideas 61
insects
    ants 16, 25
    chemical communication by 16
    echolocation by 10
    surface waves and 16
    ultrasonic hearing of 11
intention 58, 60, 103
    animal 43
    conscious 45, 48, 54, 56, 75
    definition 5
    movements 44
internal states, communication of 29,
        53

intervening variables 48
introspection 56, 64, 101

K

kidneys, physiology of 53
Krogh's principle 95

L

landmarks, uses by honeybees 67
language
  consciousness and 69
  covert 65
  criteria of 46
  design features 34 ff.
  distinguished from animal communi-
    cation 45
  held uniquely human 31 ff., 50, 96,
    103
  versus communication 49
  as window to mental experiences
    100, 105
larynx of chimpanzees 19
laughing gull communication 91
long-call, of laughing gulls 91
learning 54, 57
leks 82
linguistics, comparative 95 ff.
locomotion, coordinated 7
logical positivism 48
lying, animal 19, 45

M

magnetic monopoles 59
magnetic sensitivity
  in fish, honeybees, pigeons 12
maps, cognitive 13
materialism, dissension from 48
memory 57
mental experiences 58
  adaptive value of 82
  animal and human 3, 71, 74, 87, 98
  definition 5
  denied to animals 68, 104
  evolutionary continuity of 55
  as intervening variable 48
  nature-nurture and 61
  reality of 7
  species differences in 70

mental images 78, 90
mesons 59
metabolic energy 82
microphones, directional 10
middle concept 43, 83
mind
  analogy to computer software 7
  animal 87 ff.
  definition 5
  relation to brain 7, 53, 65
mirror images, animals and 90
mitochondria 53
models
  animal communication with 89, 90,
    105
  of honeybees 92, 94
  linguistic 92
  of living mechanisms 52
monkeys, pheromones of 16
moral values 63
  alleged threats to 46
Morgan's canon 40, 47, 71
motivation 59
Myotis lucifugus 10

N

nature-nurture question 61
nervous systems 69
neural template 58, 83, 90
neuroendocrine mechanisms 71
neurophysiology 51
  comparative 70
  mental experiences and 70
nonverbal communication 65, 88
nutrition, experiments on 96

O

observational learning of language 88
Occam's razor 47
ocelli, of honeybees 28
odors
  honeybee communication and 22
openness
  in animal communication 91
orientation
  of animals 3, 8 ff.
  electric fish 83

sun and star 9, 102
oxidation of living tissues 73

P

pain 6, 60, 78
paleontology 100
parsimony 14, 25 ff., 40, 47, 55, 60, 63
    ff.
pride of 53
participatory experiments, *see* two-way
    communication
passions
    of animals 60
    of scientists 61
pathogens 46
pattern recognition 58
    by machines 52
perceptions, animal 42
pets, communication with 89
phalangers, flying
    chemical communication of 16
pheromones 16
phototactic machines 52
phototactic orientation 9
physiological cost of display 81
pigeons
    orientation by polarized light and
        magnetism 12
plans of animals 43
playback experiments with animal sig-
    nals 88, 90
pleasure, animal 78
polarized light
    arthropods and 12
    bee orientation by 11
    bird orientation by 12
porpoises, *see* dolphins
positivism, logical 48
predation 81
predator-distraction displays 75
prevarication, animal 19, 45
"private data"
    rejected by psychologists 6
propolis, dances about 22
psychic world
    behaviorists' rejection of 64
psychoneural identity 61, 70, 98, 104
purposive behaviorism 48

Q

quality
    conveyed by bee dances 22, 93
quantum physics 100
quarks 59
quotations, annotated 4

R

rage 6
rationality, denied to bees 44
rats, maze-learning by 48
"raw feels" 6
recognition, individual vii, 15
reductionism 13, 25, 44, 60, 71, 72
reference, linguistic 41
reflex chains 85
regurgitation by bees 22
reinforcement 65
repetition rate of bat sound 10
resinous materials, honeybee dances
    about, *see* propolis
respectibility blanket 47
rigidity, assumed
    of animal behavior 67
    of animal communication 41–42, 47
ritualization of displays 75, 81
Rockefeller University, The vii
rodents, pheromones of 16
round dances of honeybees 20

S

Sarah (chimpanzee) 18
Schwanzeltanz of bees 29
Schwirrlauf of bees 24
sea otters 54
search image 58, 84
sensations in animals 60
sense-consciousness 42
sensory channels, unexpected 12
shelters, construction of 76
sign language of apes 16, 42, 89
signals, unconscious 26
signing, by anthropologists 88
signs
    distinguished from symbols 24
    use by animals 45
simplicity filters 9, 53, 56, 67, 74, 93
"so what?" objection 73

social communication 81, 96
   by electric fish 11
social organization vii
Sollwert 58, 83
sounds, of honeybees 22
specific action potential 59
speculations, importance in scientific research 4
spiders, web vibrations of 16
spontaneity 58
star-compass orientation 9
stimulus-response formulation 60
strait jacket, scientific 74
stridulation, by ants 16
subjective experiences
   in animals 49, 54, 78
subjectivity, human 64
sun-compass orientation 9
superiority, human 57
surface waves used by insects 16
surface-to-volume ratio 82
surrogates, animal 57
swarming of bees 22
symbol, definition of 24
   use of by chimpanzees 18
symbolic communication 90
   by honeybees 19, 25
   skepticism about 26
synapses 70

T

taboos, psychological 57
tactile communication 16
tactile sense organs 22
tandem running of ants 25
Tanzsprache of bees 19
template, neural 83, 90
thinking 65, 67

linked to language 97, 103
thought 58
tokens, used by chimpanzees 18
tool-making and using 54
trophallaxis in bees 22, 93
tropisms 40, 52
two-way communication 89, 97

U

understanding 58, 86
uniqueness, human 54, 63, 103
updrafts, used by gulls 78

V

verbal behavior, dominant hemisphere and 51
vibrations
   of substrate 16
   uses by honeybees 22
visual signals 16
vitalistic essences 8

W

waggle dance of honeybee 20
Washoe 42, 59, 89, 94, 97
   gestural language of 17 ff.
water, dances about 22
whales, echolocation by 11
whole-animal organization
   analogy to mind-brain relation 7
woodpecker drumming 25, 90
words
   linked to thoughts 69, 97
   recognized by chimpanzees 17

Z

Zeitgeist 58, 71

AUTHOR INDEX

A

Adams, D. K., 47, 49
Adler, M. J., 46, 57, 61
Alston, W. P., 6
Altmann, S. A., 34
Anshen, R. N., 32
Apter, M. J., 6
Armstrong, E. A., 75–76
Ayala, J., 8

B

Bastian, J. A., 27
Beer, C. G., viii, 15, 91–92
Bellugi, V., 18
Bennett, J., 44
Berg, P., 46
Bertrand, M., 49
Bever, T. G., 31, 49
Black, M., 31, 45
Blair, R., viii
Bloomfield, L., 31
Boring, E. G., 5
Boyle, D. G., 47, 48, 60
Brass, M. E., 90
Brewer, W. F., 49
Bronowski, J., 18
Brown, J. L., 24, 61
Brown, R., 41, 45
Brown, R. G. B., 75
Brown, S. R., 64
Bullock, T. H., 11, 16

C

Callender, C., 31, 82
Cassirer, E., 31
Chauvin-Muckensturm, B., 25, 26, 90
Chomsky, N., 32–34, 49, 65, 69, 96
Chown, W., 17
Cowgill, V. M., 51
Crane, J., viii, 15, 81
Creitz, L. E., 90
Critchley, M., 31, 57
Croze, H., 84

D

Daanje, A., 44
Danks, J. H., 26
Darwin, C., 98–99
Descartes, R., 32, 33, 60
Devine, J. V., 43, 83
Dobzhansky, T. H., 8, 54

E

Eccles, J. C., 8, 51, 52
Elithorn, A., 6
Emlen, S. T., 9
Esch, H., 22, 27, 92
Estes, W. K., 48

F

Falls, J. B., 15, 91
Feigl, H., 59
Feyerabend, P. K., 61
Fodor, J. A., 31, 48, 49, 59
Fouts, R. S., viii, 17, 18, 54, 56, 101
Fraenkel, G. S., 9
Frisch, K. von, 12, 19–22, 24, 27, 29, 34, 36, 42, 44, 54, 67, 74, 93–94

G

Galambos, R., 9
Galanter, E., 50
Gardner, B. T. and R. A., 15–18, 37, 55, 89, 90, 94, 101
Garrett, M. F., 31, 49
Gazzaniga, M. S., 52
Gill, T. V., 18
Glucksberg, S., 26
Goldstein, K., 31
Goodall, J. van Lawick, 16, 19, 51, 54, 95
Goodwin, L., 17
Gould, J. L., viii, 27, 28, 42, 44, 68, 92
Gould, S. J., 18, 54
Gramza, A. F., 75
Green, R. P., 12

Gregg, L. W., 7
Grice, H. P., 45
Griffin, D. R., 9, 10, 11, 13, 18
Gunn, D. L., 9

H

Hailman, J. P., 39
Halperin, S., 19
Hamilton, W. H., 90
Hampshire, S., 43, 44
Hartridge, H., 10
Hartshorne, C., 77
Hattiangadi, J. M., 31
Healy, A. F., 31
Hebb, D. O., 5, 60, 64
Henery, J., 27
Hewes, G. W., 88
Hinde, R. A., 15, 26, 88
Hockett, C. F., 34
Holloway, R. L., 56
Hopkins, C. D., 16, 83, 88, 91
Howell, D. J., 11, 13
Hubbard, J. I., 8
Hutchinson, G. E., 69
Huxley, J., 82–83

I

Irwin, F. W., 49

J

Jennings, H. S., 47, 59, 74
Jones, D., 7
Jordan, H., viii

K

Kalmijn, A. J., 12
Keeton, W. T., 12
Kellogg, W. N., 11
Kenny, A. J. P., 42
Kimble, G. A., 49
Klein, D. B., 39
Klima, E. S., 18
Klopfer, P. H., 39
Kramer, G., 9
Krames, L., 88

Krebs, H. A., 95
Krebs, J. R., 91
Kreithen, M. L., 12
Krogh, A., 95

L

Langer, S. K., 26, 29, 31, 40, 50, 57
Lashley, K. S., 5, 39, 56, 63–66
Lehrer, K., 59
Lemmon, W., 17
Lenneberg, E. H., 31
Lindauer, M., 12, 22–24, 27, 54, 55, 59
Linden, E., 18, 54
Lissmann, H. W., 11
Lloyd, J. E., 16
Loeb, J., 39, 52, 60
Longuet-Higgins, H. C., 42, 43
Lorenz, K., 8, 34, 39, 49, 59, 78
Lucas, J. R., 42

M

Machin, K. E., 11
MacLeod, M. C., 27
Malcolm, N., 60
Maritain, J., 45, 61
Marler, P., vii, viii, 16, 47, 83, 90, 98
Marshall, A. J., 76–77
Martin, H., 12
Matthews, G. V. T., 9
Maxwell, G., 61
McGrew, W. C., 54
McGuigan, F. J., 66
McMullan, E. M., 45
Menzel, E. W., Jr., 19
Michael, C., 10
Michener, C. D., 29
Miller, G. A., 50, 71–72
Mittelstaedt, H., 83
Moehres, F. P., 13
Monod, J. L., 31
Morgan, C. L., 39, 40, 47, 71–72

Morris, C., 24, 25
Mowrer, O. H., 48, 49, 57, 68

N

Nagel, T., vii
Nance, J., 88
Nebes, R. D., 51
Neisser, U., 48
Norris, K. S., 11
Nottebohm, F., vii

O

Oettingen-Spielberg, T. von, 13

P

Perlmuter, L. C., 49
Pfungst, O., 26
Piaget, J., 8
Pierce, G. W., 9, 10
Pollio, H. R., 33, 61
Polten, E. P., 59
Popper, K. R., 8, 40, 50
Premack, D., 18, 37, 98
Pribram, K. H., 50, 52
Price, H. H., 33
Pyles, T., 31

R

Ramsay, A., 15
Razran, G., 49
Rensch, B., 50
Rigby, R. L., 17, 18
Robinson, D. N., 45
Roeder, K., 11
Rohles, F. H., 43, 83
Rumbaugh, D. M., viii, 18

S

Sarles, H., 34
Sauer, E. G. F., 9
Schmidt-Koenig, K., 9
Schoonover, R. A., 66
Schultz, D., 39, 48
Scriven, M., 6, 8
Sebeok, T. A., 15, 16
Sellars, W., 59
Simmons, J. A., 11, 13

Simpson, G. G., 54
Skinner, B. F., 64–67, 84
Skutch, A. F., 75
"Smith, Adam," 5, 56
Smith, K., 8
Smith, W. J., viii, 5, 8, 15, 29. 53. 91
Sperry, R. W., 8, 51, 53
Stenhouse, B., 39
Stevenson, J. C., 91
Stout, J. F., 90
Suga, N., 11, 13
Suthers, R. A., 11

T

Tavolga, W. N., 26
Tax, S., 82
Taylor, J. G., 48
Teng, E. L., 51
Terwilliger, R. F., 67–68
Thass-Thienemann, T., 31
Thomas, L., 44
Thorpe, W. H., 13, 18, 34, 37, 59
Tinbergen, N., 44, 82, 90
Tolman, E. C., 6, 13, 48
Treat, A., 11

W

Waddington, C. H., 42
Walcott, C., 12
Watson, J. B., 39, 60, 64, 68
Webster, F. A., 10
Weiner, B., 49
Weis, D. D., 31
Wells, P. H., 26
Wenner, A. M., 22, 26, 29
Westby, G. W. M., 16, 91
Whitehead, A. N., 37
Wilcox, C. R., 90
Wilcox, R. S., 16
Wilson, E. O., 16, 25, 29, 75. 82, 85, 91
Wiltschko, W., 12
Wittgenstein, L., 50, 68, 97

Y

Yankelovich, D., 56